Activities Workbook
for
Active Calculus Single Variable
Chapters 5–8

Matthew Boelkins
Grand Valley State University

Contributing Authors

David Austin
Grand Valley State University

Steven Schlicker
Grand Valley State University

Production Editor

Mitchel T. Keller
Morningside College

July 26, 2019

Cover Photo: James Haefner Photography

Edition: 2018 Updated

Website: http://activecalculus.org

Preface

This Activities Workbook for *Active Calculus Single Variable* collects all the Preview Activities and Activities in a way that each starts on a new page. The Activities Workbook is designed to be used by students who wish to have a complete set of the activities to work through as they read the book in an electronic format and for instructors who wish to have a one activity per page format to make printing for distribution in class easier.

The design of this workbook is such that each Preview Activity and Activity starts on a right-hand page. As a result, most left-hand pages in this workbook are intentionally left blank as a place for student work associated with one of the adjacent activities. The workbook is offered for purchase in print form in two volumes: the first for chapters 1–4 and the second for chapters 5–8.

Contents

Contents

CHAPTER 5

Evaluating Integrals

5.1 Constructing Accurate Graphs of Antiderivatives

Preview Activity 5.1.1. Suppose that the following information is known about a function f: the graph of its derivative, $y = f'(x)$, is given in Figure 5.1.1. Further, assume that f' is piecewise linear (as pictured) and that for $x \leq 0$ and $x \geq 6$, $f'(x) = 0$. Finally, it is given that $f(0) = 1$.

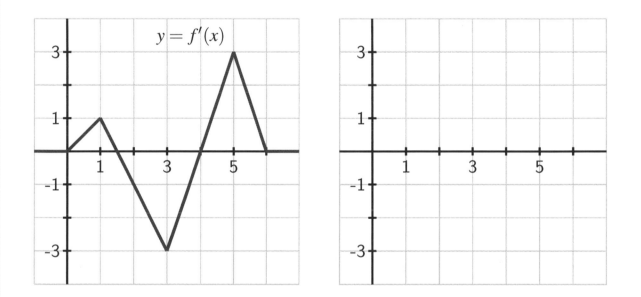

Figure 5.1.1: At left, the graph of $y = f'(x)$; at right, axes for plotting $y = f(x)$.

a. On what interval(s) is f an increasing function? On what intervals is f decreasing?

b. On what interval(s) is f concave up? concave down?

c. At what point(s) does f have a relative minimum? a relative maximum?

d. Recall that the Total Change Theorem tells us that

$$f(1) - f(0) = \int_0^1 f'(x)\,dx.$$

What is the exact value of $f(1)$?

e. Use the given information and similar reasoning to that in (d) to determine the exact value of $f(2)$, $f(3)$, $f(4)$, $f(5)$, and $f(6)$.

f. Based on your responses to all of the preceding questions, sketch a complete and accurate graph of $y = f(x)$ on the axes provided, being sure to indicate the behavior of f for $x < 0$ and $x > 6$.

Activity 5.1.2. Suppose that the function $y = f(x)$ is given by the graph shown in Figure 5.1.2, and that the pieces of f are either portions of lines or portions of circles. In addition, let F be an antiderivative of f and say that $F(0) = -1$. Finally, assume that for $x \le 0$ and $x \ge 7$, $f(x) = 0$.

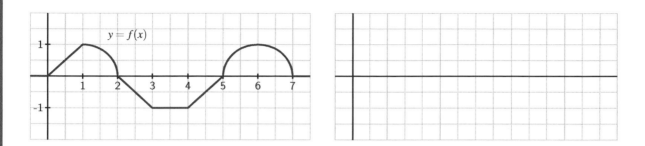

Figure 5.1.2: At left, the graph of $y = f(x)$.

a. On what interval(s) is F an increasing function? On what intervals is F decreasing?

b. On what interval(s) is F concave up? concave down? neither?

c. At what point(s) does F have a relative minimum? a relative maximum?

d. Use the given information to determine the exact value of $F(x)$ for $x = 1, 2, \ldots, 7$. In addition, what are the values of $F(-1)$ and $F(8)$?

e. Based on your responses to all of the preceding questions, sketch a complete and accurate graph of $y = F(x)$ on the axes provided, being sure to indicate the behavior of F for $x < 0$ and $x > 7$. Clearly indicate the scale on the vertical and horizontal axes of your graph.

f. What happens if we change one key piece of information: in particular, say that G is an antiderivative of f and $G(0) = 0$. How (if at all) would your answers to the preceding questions change? Sketch a graph of G on the same axes as the graph of F you constructed in (e).

3

Activity 5.1.3. For each of the following functions, sketch an accurate graph of the antiderivative that satisfies the given initial condition. In addition, sketch the graph of two additional antiderivatives of the given function, and state the corresponding initial conditions that each of them satisfy. If possible, find an algebraic formula for the antiderivative that satisfies the initial condition.

a. original function: $g(x) = |x| - 1$; initial condition: $G(-1) = 0$; interval for sketch: $[-2, 2]$

b. original function: $h(x) = \sin(x)$; initial condition: $H(0) = 1$; interval for sketch: $[0, 4\pi]$

c. original function: $p(x) = \begin{cases} x^2, & \text{if } 0 < x < 1 \\ -(x-2)^2, & \text{if } 1 < x < 2 \\ 0 & \text{otherwise} \end{cases}$; initial condition: $P(0) = 1$; interval for sketch: $[-1, 3]$

Activity 5.1.4. Suppose that g is given by the graph at left in Figure 5.1.6 and that A is the corresponding integral function defined by $A(x) = \int_1^x g(t)\,dt$.

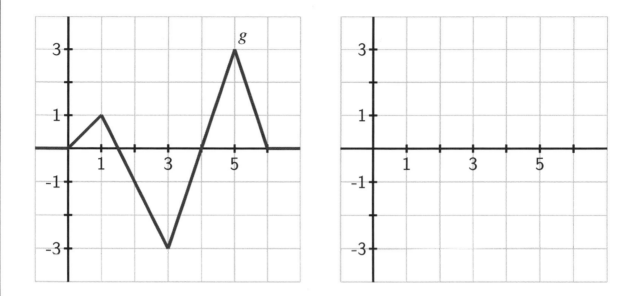

Figure 5.1.6: At left, the graph of $y = g(t)$; at right, axes for plotting $y = A(x)$, where A is defined by the formula $A(x) = \int_1^x g(t)\,dt$.

a. On what interval(s) is A an increasing function? On what intervals is A decreasing? Why?

b. On what interval(s) do you think A is concave up? concave down? Why?

c. At what point(s) does A have a relative minimum? a relative maximum?

d. Use the given information to determine the exact values of $A(0)$, $A(1)$, $A(2)$, $A(3)$, $A(4)$, $A(5)$, and $A(6)$.

e. Based on your responses to all of the preceding questions, sketch a complete and accurate graph of $y = A(x)$ on the axes provided, being sure to indicate the behavior of A for $x < 0$ and $x > 6$.

f. How does the graph of B compare to A if B is instead defined by $B(x) = \int_0^x g(t)\,dt$?

5.2 The Second Fundamental Theorem of Calculus

Preview Activity 5.2.1. Consider the function A defined by the rule

$$A(x) = \int_1^x f(t)\, dt,$$

where $f(t) = 4 - 2t$.

a. Compute $A(1)$ and $A(2)$ exactly.

b. Use the First Fundamental Theorem of Calculus to find a formula for $A(x)$ that does not involve integrals. That is, use the first FTC to evaluate $\int_1^x (4 - 2t)\, dt$.

c. Observe that f is a linear function; what kind of function is A?

d. Using the formula you found in (b) that does not involve integrals, compute $A'(x)$.

e. While we have defined f by the rule $f(t) = 4 - 2t$, it is equivalent to say that f is given by the rule $f(x) = 4 - 2x$. What do you observe about the relationship between A and f?

Activity 5.2.2. Suppose that f is the function given in Figure 5.2.2 and that f is a piecewise function whose parts are either portions of lines or portions of circles, as pictured.

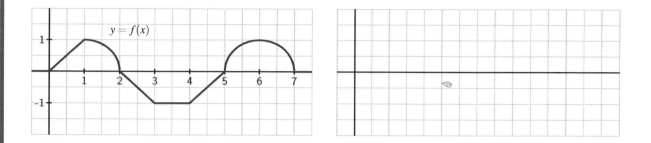

Figure 5.2.2: At left, the graph of $y = f(x)$. At right, axes for sketching $y = A(x)$.

In addition, let A be the function defined by the rule $A(x) = \int_2^x f(t)\,dt$.

a. What does the Second FTC tell us about the relationship between A and f?

b. Compute $A(1)$ and $A(3)$ exactly.

c. Sketch a precise graph of $y = A(x)$ on the axes at right that accurately reflects where A is increasing and decreasing, where A is concave up and concave down, and the exact values of A at $x = 0, 1, \ldots, 7$.

d. How is A similar to, but different from, the function F that you found in Activity 5.1.2?

e. With as little additional work as possible, sketch precise graphs of the functions $B(x) = \int_3^x f(t)\,dt$ and $C(x) = \int_1^x f(t)\,dt$. Justify your results with at least one sentence of explanation.

Activity 5.2.3. Suppose that $f(t) = \frac{t}{1+t^2}$ and $F(x) = \int_0^x f(t)\,dt$.

a. On the axes at left in Figure 5.2.5, plot a graph of $f(t) = \frac{t}{1+t^2}$ on the interval $-10 \le t \le 10$. Clearly label the vertical axes with appropriate scale.

b. What is the key relationship between F and f, according to the Second FTC?

c. Use the first derivative test to determine the intervals on which F is increasing and decreasing.

d. Use the second derivative test to determine the intervals on which F is concave up and concave down. Note that $f'(t)$ can be simplified to be written in the form $f'(t) = \frac{1-t^2}{(1+t^2)^2}$.

e. Using technology appropriately, estimate the values of $F(5)$ and $F(10)$ through appropriate Riemann sums.

f. Sketch an accurate graph of $y = F(x)$ on the righthand axes provided, and clearly label the vertical axes with appropriate scale.

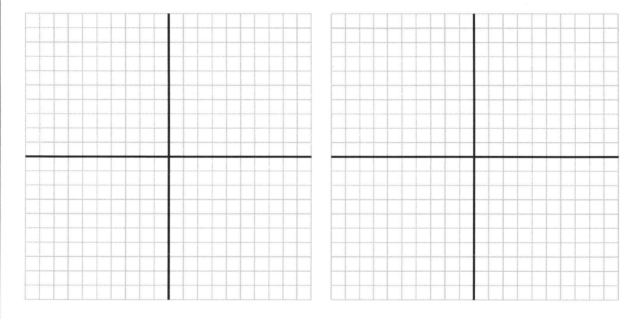

Figure 5.2.5: Axes for plotting f and F.

Activity 5.2.4. Evaluate each of the following derivatives and definite integrals. Clearly cite whether you use the First or Second FTC in so doing.

a. $\frac{d}{dx}\left[\int_4^x e^{t^2}\,dt\right]$

b. $\int_{-2}^x \frac{d}{dt}\left[\frac{t^4}{1+t^4}\right]\,dt$

c. $\frac{d}{dx}\left[\int_x^1 \cos(t^3)\,dt\right]$

d. $\int_3^x \frac{d}{dt}\left[\ln(1+t^2)\right]\,dt$

e. $\frac{d}{dx}\left[\int_4^{x^3} \sin(t^2)\,dt\right]$.

5.3 Integration by Substitution

Preview Activity 5.3.1. In ((((Unresolved xref, reference "sec-2-5-chain"; check spelling or use "provisional" attribute)))Section , we learned the Chain Rule and how it can be applied to find the derivative of a composite function. In particular, if u is a differentiable function of x, and f is a differentiable function of $u(x)$, then

$$\frac{d}{dx}\left[f(u(x))\right] = f'(u(x)) \cdot u'(x).$$

In words, we say that the derivative of a composite function $c(x) = f(u(x))$, where f is considered the "outer" function and u the "inner" function, is "the derivative of the outer function, evaluated at the inner function, times the derivative of the inner function."

    a. For each of the following functions, use the Chain Rule to find the function's derivative. Be sure to label each derivative by name (e.g., the derivative of $g(x)$ should be labeled $g'(x)$).

       i. $g(x) = e^{3x}$ iv. $q(x) = (2 - 7x)^4$

       ii. $h(x) = \sin(5x + 1)$

       iii. $p(x) = \arctan(2x)$ v. $r(x) = 3^{4-11x}$

    b. For each of the following functions, use your work in (a) to help you determine the general antiderivative[1] of the function. Label each antiderivative by name (e.g., the antiderivative of m should be called M). In addition, check your work by computing the derivative of each proposed antiderivative.

       i. $m(x) = e^{3x}$ iv. $v(x) = (2 - 7x)^3$

       ii. $n(x) = \cos(5x + 1)$

       iii. $s(x) = \frac{1}{1+4x^2}$ v. $w(x) = 3^{4-11x}$

    c. Based on your experience in parts (a) and (b), conjecture an antiderivative for each of the following functions. Test your conjectures by computing the derivative of each proposed antiderivative.

       i. $a(x) = \cos(\pi x)$ iii. $c(x) = xe^{x^2}$

       ii. $b(x) = (4x + 7)^{11}$

---

[1]Recall that the general antiderivative of a function includes "+C" to reflect the entire family of functions that share the same derivative.

Activity 5.3.2. Evaluate each of the following indefinite integrals. Check each antiderivative that you find by differentiating.

a. $\int \sin(8 - 3x)\, dx$

b. $\int \sec^2(4x)\, dx$

c. $\int \frac{1}{11x - 9}\, dx$

d. $\int \csc(2x + 1)\cot(2x + 1)\, dx$

e. $\int \frac{1}{\sqrt{1 - 16x^2}}\, dx$

f. $\int 5^{-x}\, dx$

Activity 5.3.3. Evaluate each of the following indefinite integrals by using these steps:

- Find two functions within the integrand that form (up to a possible missing constant) a function-derivative pair;

- Make a substitution and convert the integral to one involving u and du;

- Evaluate the new integral in u;

- Convert the resulting function of u back to a function of x by using your earlier substitution;

- Check your work by differentiating the function of x. You should come up with the integrand originally given.

a. $\int \frac{x^2}{5x^3+1}\, dx$

b. $\int e^x \sin(e^x)\, dx$

c. $\int \frac{\cos(\sqrt{x})}{\sqrt{x}}\, dx$

Activity 5.3.4. Evaluate each of the following definite integrals exactly through an appropriate u-substitution.

a. $\int_1^2 \frac{x}{1+4x^2}\, dx$

c. $\int_{2/\pi}^{4/\pi} \frac{\cos\left(\frac{1}{x}\right)}{x^2}\, dx$

b. $\int_0^1 e^{-x}(2e^{-x}+3)^9\, dx$

5.4 Integration by Parts

Preview Activity 5.4.1. In (((Unresolved xref, reference "sec-2-3-prod-quot"; check spelling or use "provisional" attribute)))Section , we developed the Product Rule and studied how it is employed to differentiate a product of two functions. In particular, recall that if f and g are differentiable functions of x, then

$$\frac{d}{dx}\left[f(x) \cdot g(x)\right] = f(x) \cdot g'(x) + g(x) \cdot f'(x).$$

a. For each of the following functions, use the Product Rule to find the function's derivative. Be sure to label each derivative by name (e.g., the derivative of $g(x)$ should be labeled $g'(x)$).

    i. $g(x) = x\sin(x)$ iv. $q(x) = x^2\cos(x)$

    ii. $h(x) = xe^x$

    iii. $p(x) = x\ln(x)$ v. $r(x) = e^x\sin(x)$

b. Use your work in (a) to help you evaluate the following indefinite integrals. Use differentiation to check your work.

    i. $\int xe^x + e^x\,dx$ iv. $\int x\cos(x) + \sin(x)\,dx$

    ii. $\int e^x(\sin(x) + \cos(x))\,dx$

    iii. $\int 2x\cos(x) - x^2\sin(x)\,dx$ v. $\int 1 + \ln(x)\,dx$

c. Observe that the examples in (b) work nicely because of the derivatives you were asked to calculate in (a). Each integrand in (b) is precisely the result of differentiating one of the products of basic functions found in (a). To see what happens when an integrand is still a product but not necessarily the result of differentiating an elementary product, we consider how to evaluate

$$\int x\cos(x)\,dx.$$

    i. First, observe that

$$\frac{d}{dx}[x\sin(x)] = x\cos(x) + \sin(x).$$

    Integrating both sides indefinitely and using the fact that the integral of a sum is the sum of the integrals, we find that

$$\int\left(\frac{d}{dx}[x\sin(x)]\right)dx = \int x\cos(x)\,dx + \int \sin(x)\,dx.$$

    In this last equation, evaluate the indefinite integral on the left side as well as the rightmost indefinite integral on the right.

    ii. In the most recent equation from (i.), solve the equation for the expression $\int x\cos(x)\,dx$.

    iii. For which product of basic functions have you now found the antiderivative?

Activity 5.4.2. Evaluate each of the following indefinite integrals. Check each antiderivative that you find by differentiating.

a. $\int te^{-t}\,dt$

c. $\int z\sec^2(z)\,dz$

b. $\int 4x\sin(3x)\,dx$

d. $\int x\ln(x)\,dx$

Activity 5.4.3. Evaluate each of the following indefinite integrals, using the provided hints.

a. Evaluate $\int \arctan(x)\, dx$ by using Integration by Parts with the substitution $u = \arctan(x)$ and $dv = 1\, dx$.

b. Evaluate $\int \ln(z)\, dz$. Consider a similar substitution to the one in (a).

c. Use the substitution $z = t^2$ to transform the integral $\int t^3 \sin(t^2)\, dt$ to a new integral in the variable z, and evaluate that new integral by parts.

d. Evaluate $\int s^5 e^{s^3}\, ds$ using an approach similar to that described in (c).

e. Evaluate $\int e^{2t} \cos(e^t)\, dt$. You will find it helpful to note that $e^{2t} = e^t \cdot e^t$.

Activity 5.4.4. Evaluate each of the following indefinite integrals.

a. $\int x^2 \sin(x)\, dx$

b. $\int t^3 \ln(t)\, dt$

c. $\int e^z \sin(z)\, dz$

d. $\int s^2 e^{3s}\, ds$

e. $\int t \arctan(t)\, dt$ (*Hint:* At a certain point in this problem, it is very helpful to note that $\frac{t^2}{1+t^2} = 1 - \frac{1}{1+t^2}$.)

5.5 Other Options for Finding Algebraic Antiderivatives

Preview Activity 5.5.1. For each of the indefinite integrals below, the main question is to decide whether the integral can be evaluated using u-substitution, integration by parts, a combination of the two, or neither. For integrals for which your answer is affirmative, state the substitution(s) you would use. It is not necessary to actually evaluate any of the integrals completely, unless the integral can be evaluated immediately using a familiar basic antiderivative.

a. $\int x^2 \sin(x^3)\,dx$, $\int x^2 \sin(x)\,dx$, $\int \sin(x^3)\,dx$, $\int x^5 \sin(x^3)\,dx$

b. $\int \frac{1}{1+x^2}\,dx$, $\int \frac{x}{1+x^2}\,dx$, $\int \frac{2x+3}{1+x^2}\,dx$, $\int \frac{e^x}{1+(e^x)^2}\,dx$,

c. $\int x \ln(x)\,dx$, $\int \frac{\ln(x)}{x}\,dx$, $\int \ln(1+x^2)\,dx$, $\int x \ln(1+x^2)\,dx$,

d. $\int x\sqrt{1-x^2}\,dx$, $\int \frac{1}{\sqrt{1-x^2}}\,dx$, $\int \frac{x}{\sqrt{1-x^2}}\,dx$, $\int \frac{1}{x\sqrt{1-x^2}}\,dx$,

Activity 5.5.2. For each of the following problems, evaluate the integral by using the partial fraction decomposition provided.

a. $\int \frac{1}{x^2-2x-3} \, dx$, given that $\frac{1}{x^2-2x-3} = \frac{1/4}{x-3} - \frac{1/4}{x+1}$

b. $\int \frac{x^2+1}{x^3-x^2} \, dx$, given that $\frac{x^2+1}{x^3-x^2} = -\frac{1}{x} - \frac{1}{x^2} + \frac{2}{x-1}$

c. $\int \frac{x-2}{x^4+x^2} \, dx$, given that $\frac{x-2}{x^4+x^2} = \frac{1}{x} - \frac{2}{x^2} + \frac{-x+2}{1+x^2}$

Activity 5.5.3. For each of the following integrals, evaluate the integral using u-substitution and/or an entry from the table found in Appendix A.

a. $\int \sqrt{x^2 + 4}\, dx$

b. $\int \frac{x}{\sqrt{x^2+4}}\, dx$

c. $\int \frac{2}{\sqrt{16+25x^2}}\, dx$

d. $\int \frac{1}{x^2\sqrt{49-36x^2}}\, dx$

5.6 Numerical Integration

Preview Activity 5.6.1. As we begin to investigate ways to approximate definite integrals, it will be insightful to compare results to integrals whose exact values we know. To that end, the following sequence of questions centers on $\int_0^3 x^2 \, dx$.

a. Use the applet at http://gvsu.edu/s/a9 with the function $f(x) = x^2$ on the window of x values from 0 to 3 to compute L_3, the left Riemann sum with three subintervals.

b. Likewise, use the applet to compute R_3 and M_3, the right and middle Riemann sums with three subintervals, respectively.

c. Use the Fundamental Theorem of Calculus to compute the exact value of $I = \int_0^3 x^2 \, dx$.

d. We define the *error* that results from an approximation of a definite integral to be the approximation's value minus the integral's exact value. What is the error that results from using L_3? From R_3? From M_3?

e. In what follows in this section, we will learn a new approach to estimating the value of a definite integral known as the Trapezoid Rule. The basic idea is to use trapezoids, rather than rectangles, to estimate the area under a curve. What is the formula for the area of a trapezoid with bases of length b_1 and b_2 and height h?

f. Working by hand, estimate the area under $f(x) = x^2$ on $[0,3]$ using three subintervals and three corresponding trapezoids. What is the error in this approximation? How does it compare to the errors you calculated in (d)?

Activity 5.6.2. In this activity, we explore the relationships among the errors generated by left, right, midpoint, and trapezoid approximations to the definite integral $\int_1^2 \frac{1}{x^2} \, dx$.

a. Use the First FTC to evaluate $\int_1^2 \frac{1}{x^2} \, dx$ exactly.

b. Use appropriate computing technology to compute the following approximations for $\int_1^2 \frac{1}{x^2} \, dx$: T_4, M_4, T_8, and M_8.

c. Let the *error* that results from an approximation be the approximation's value minus the exact value of the definite integral. For instance, if we let $E_{T,4}$ represent the error that results from using the trapezoid rule with 4 subintervals to estimate the integral, we have

$$E_{T,4} = T_4 - \int_1^2 \frac{1}{x^2} \, dx.$$

Similarly, we compute the error of the midpoint rule approximation with 8 subintervals by the formula

$$E_{M,8} = M_8 - \int_1^2 \frac{1}{x^2} \, dx.$$

Based on your work in (a) and (b) above, compute $E_{T,4}$, $E_{T,8}$, $E_{M,4}$, $E_{M,8}$.

d. Which rule consistently over-estimates the exact value of the definite integral? Which rule consistently under-estimates the definite integral?

e. What behavior(s) of the function $f(x) = \frac{1}{x^2}$ lead to your observations in (d)?

Activity 5.6.3. A car traveling along a straight road is braking and its velocity is measured at several different points in time, as given in the following table. Assume that v is continuous, always decreasing, and always decreasing at a decreasing rate, as is suggested by the data.

seconds, t	Velocity in ft/sec, $v(t)$
0	100
0.3	99
0.6	96
0.9	90
1.2	80
1.5	50
1.8	0

Table 5.6.7: Data for the braking car.

Figure 5.6.8: Axes for plotting the data in Activity 5.6.3.

a. Plot the given data on the set of axes provided in Figure 5.6.8 with time on the horizontal axis and the velocity on the vertical axis.

b. What definite integral will give you the exact distance the car traveled on $[0, 1.8]$?

c. Estimate the total distance traveled on $[0, 1.8]$ by computing L_3, R_3, and T_3. Which of these under-estimates the true distance traveled?

d. Estimate the total distance traveled on $[0, 1.8]$ by computing M_3. Is this an over- or under-estimate? Why?

e. Using your results from (c) and (d), improve your estimate further by using Simpson's Rule.

f. What is your best estimate of the average velocity of the car on $[0, 1.8]$? Why? What are the units on this quantity?

Activity 5.6.4. Consider the functions $f(x) = 2-x^2$, $g(x) = 2-x^3$, and $h(x) = 2-x^4$, all on the interval $[0,1]$. For each of the questions that require a numerical answer in what follows, write your answer exactly in fraction form.

a. On the three sets of axes provided in Figure 5.6.9, sketch a graph of each function on the interval $[0,1]$, and compute L_1 and R_1 for each. What do you observe?

b. Compute M_1 for each function to approximate $\int_0^1 f(x)\,dx$, $\int_0^1 g(x)\,dx$, and $\int_0^1 h(x)\,dx$, respectively.

c. Compute T_1 for each of the three functions, and hence compute S_2 for each of the three functions.

d. Evaluate each of the integrals $\int_0^1 f(x)\,dx$, $\int_0^1 g(x)\,dx$, and $\int_0^1 h(x)\,dx$ exactly using the First FTC.

e. For each of the three functions f, g, and h, compare the results of L_1, R_1, M_1, T_1, and S_2 to the true value of the corresponding definite integral. What patterns do you observe?

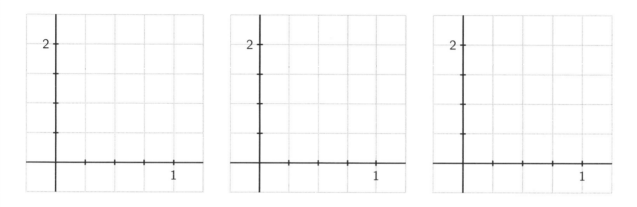

Figure 5.6.9: Axes for plotting the functions in Activity 5.6.4.

CHAPTER **6**

Using Definite Integrals

6.1 Using Definite Integrals to Find Area and Length

Preview Activity 6.1.1. Consider the functions given by $f(x) = 5 - (x - 1)^2$ and $g(x) = 4 - x$.

a. Use algebra to find the points where the graphs of f and g intersect.

b. Sketch an accurate graph of f and g on the axes provided, labeling the curves by name and the intersection points with ordered pairs.

c. Find and evaluate exactly an integral expression that represents the area between $y = f(x)$ and the x-axis on the interval between the intersection points of f and g.

d. Find and evaluate exactly an integral expression that represents the area between $y = g(x)$ and the x-axis on the interval between the intersection points of f and g.

e. What is the exact area between f and g between their intersection points? Why?

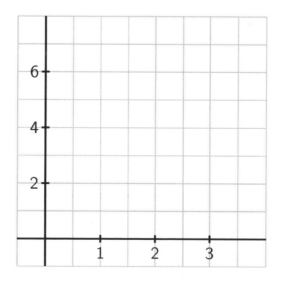

Figure 6.1.1: Axes for plotting f and g in Preview Activity 6.1.1

Activity 6.1.2. In each of the following problems, our goal is to determine the area of the region described. For each region, (i) determine the intersection points of the curves, (ii) sketch the region whose area is being found, (iii) draw and label a representative slice, and (iv) state the area of the representative slice. Then, state a definite integral whose value is the exact area of the region, and evaluate the integral to find the numeric value of the region's area.

a. The finite region bounded by $y = \sqrt{x}$ and $y = \frac{1}{4}x$.

b. The finite region bounded by $y = 12 - 2x^2$ and $y = x^2 - 8$.

c. The area bounded by the y-axis, $f(x) = \cos(x)$, and $g(x) = \sin(x)$, where we consider the region formed by the first positive value of x for which f and g intersect.

d. The finite regions between the curves $y = x^3 - x$ and $y = x^2$.

Activity 6.1.3. In each of the following problems, our goal is to determine the area of the region described. For each region, (i) determine the intersection points of the curves, (ii) sketch the region whose area is being found, (iii) draw and label a representative slice, and (iv) state the area of the representative slice. Then, state a definite integral whose value is the exact area of the region, and evaluate the integral to find the numeric value of the region's area. *Note well:* At the step where you draw a representative slice, you need to make a choice about whether to slice vertically or horizontally.

    a. The finite region bounded by $x = y^2$ and $x = 6 - 2y^2$.

    b. The finite region bounded by $x = 1 - y^2$ and $x = 2 - 2y^2$.

    c. The area bounded by the x-axis, $y = x^2$, and $y = 2 - x$.

    d. The finite regions between the curves $x = y^2 - 2y$ and $y = x$.

Activity 6.1.4. Each of the following questions somehow involves the arc length along a curve.

  a. Use the definition and appropriate computational technology to determine the arc length along $y = x^2$ from $x = -1$ to $x = 1$.

  b. Find the arc length of $y = \sqrt{4 - x^2}$ on the interval $-2 \le x \le 2$. Find this value in two different ways: (a) by using a definite integral, and (b) by using a familiar property of the curve.

  c. Determine the arc length of $y = xe^{3x}$ on the interval $[0, 1]$.

  d. Will the integrals that arise calculating arc length typically be ones that we can evaluate exactly using the First FTC, or ones that we need to approximate? Why?

  e. A moving particle is traveling along the curve given by $y = f(x) = 0.1x^2 + 1$, and does so at a constant rate of 7 cm/sec, where both x and y are measured in cm (that is, the curve $y = f(x)$ is the path along which the object actually travels; the curve is not a "position function"). Find the position of the particle when $t = 4$ sec, assuming that when $t = 0$, the particle's location is $(0, f(0))$.

---

[1]This integral is actually "improper" because the integrand is undefined at the endpoints, $x = \pm 2$. We will learn how to evaluate such integrals in Section 6.5.

6.2 Using Definite Integrals to Find Volume

Preview Activity 6.2.1. Consider a circular cone of radius 3 and height 5, which we view horizontally as pictured in Figure 6.2.1. Our goal in this activity is to use a definite integral to determine the volume of the cone.

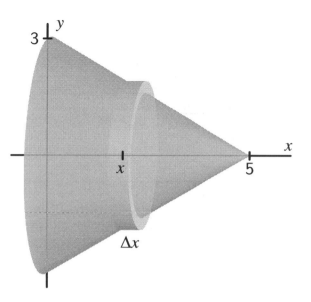

Figure 6.2.1: The circular cone described in Preview Activity 6.2.1

a. Find a formula for the linear function $y = f(x)$ that is pictured in Figure 6.2.1.

b. For the representative slice of thickness Δx that is located horizontally at a location x (somewhere between $x = 0$ and $x = 5$), what is the radius of the representative slice? Note that the radius depends on the value of x.

c. What is the volume of the representative slice you found in (b)?

d. What definite integral will sum the volumes of the thin slices across the full horizontal span of the cone? What is the exact value of this definite integral?

e. Compare the result of your work in (d) to the volume of the cone that comes from using the formula $V_{\text{cone}} = \frac{1}{3}\pi r^2 h$.

Activity 6.2.2. In each of the following questions, draw a careful, labeled sketch of the region described, as well as the resulting solid that results from revolving the region about the stated axis. In addition, draw a representative slice and state the volume of that slice, along with a definite integral whose value is the volume of the entire solid. It is not necessary to evaluate the integrals you find.

a. The region S bounded by the x-axis, the curve $y = \sqrt{x}$, and the line $x = 4$; revolve S about the x-axis.

b. The region S bounded by the y-axis, the curve $y = \sqrt{x}$, and the line $y = 2$; revolve S about the x-axis.

c. The finite region S bounded by the curves $y = \sqrt{x}$ and $y = x^3$; revolve S about the x-axis.

d. The finite region S bounded by the curves $y = 2x^2 + 1$ and $y = x^2 + 4$; revolve S about the x-axis.

e. The region S bounded by the y-axis, the curve $y = \sqrt{x}$, and the line $y = 2$; revolve S about the y-axis. How is this problem different from the one posed in part (b)?

Activity 6.2.3. In each of the following questions, draw a careful, labeled sketch of the region described, as well as the resulting solid that results from revolving the region about the stated axis. In addition, draw a representative slice and state the volume of that slice, along with a definite integral whose value is the volume of the entire solid. It is not necessary to evaluate the integrals you find.

a. The region S bounded by the y-axis, the curve $y = \sqrt{x}$, and the line $y = 2$; revolve S about the y-axis.

b. The region S bounded by the x-axis, the curve $y = \sqrt{x}$, and the line $x = 4$; revolve S about the y-axis.

c. The finite region S in the first quadrant bounded by the curves $y = 2x$ and $y = x^3$; revolve S about the x-axis.

d. The finite region S in the first quadrant bounded by the curves $y = 2x$ and $y = x^3$; revolve S about the y-axis.

e. The finite region S bounded by the curves $x = (y-1)^2$ and $y = x - 1$; revolve S about the y-axis

Activity 6.2.4. In each of the following questions, draw a careful, labeled sketch of the region described, as well as the resulting solid that results from revolving the region about the stated axis. In addition, draw a representative slice and state the volume of that slice, along with a definite integral whose value is the volume of the entire solid. It is not necessary to evaluate the integrals you find. For each prompt, use the finite region S in the first quadrant bounded by the curves $y = 2x$ and $y = x^3$.

a. Revolve S about the line $y = -2$.

b. Revolve S about the line $y = 4$.

c. Revolve S about the line $x = -1$.

d. Revolve S about the line $x = 5$.

6.3 Density, Mass, and Center of Mass

Preview Activity 6.3.1. In each of the following scenarios, we consider the distribution of a quantity along an axis.

a. Suppose that the function $c(x) = 200 + 100e^{-0.1x}$ models the density of traffic on a straight road, measured in cars per mile, where x is number of miles east of a major interchange, and consider the definite integral $\int_0^2 (200 + 100e^{-0.1x})\,dx$.

   i. What are the units on the product $c(x) \cdot \Delta x$?

   ii. What are the units on the definite integral and its Riemann sum approximation given by

$$\int_0^2 c(x)\,dx \approx \sum_{i=1}^{n} c(x_i)\Delta x?$$

   iii. Evaluate the definite integral $\int_0^2 c(x)\,dx = \int_0^2 \left(200 + 100e^{-0.1x}\right)\,dx$ and write one sentence to explain the meaning of the value you find.

b. On a 6 foot long shelf filled with books, the function B models the distribution of the weight of the books, in pounds per inch, where x is the number of inches from the left end of the bookshelf. Let $B(x)$ be given by the rule $B(x) = 0.5 + \frac{1}{(x+1)^2}$.

   i. What are the units on the product $B(x) \cdot \Delta x$?

   ii. What are the units on the definite integral and its Riemann sum approximation given by

$$\int_{12}^{36} B(x)\,dx \approx \sum_{i=1}^{n} B(x_i)\Delta x?$$

   iii. Evaluate the definite integral $\int_0^{72} B(x)\,dx = \int_0^{72} \left(0.5 + \frac{1}{(x+1)^2}\right)\,dx$ and write one sentence to explain the meaning of the value you find.

Activity 6.3.2. Consider the following situations in which mass is distributed in a non-constant manner.

a. Suppose that a thin rod with constant cross-sectional area of 1 cm^2 has its mass distributed according to the density function $\rho(x) = 2e^{-0.2x}$, where x is the distance in cm from the left end of the rod, and the units on $\rho(x)$ are g/cm. If the rod is 10 cm long, determine the exact mass of the rod.

b. Consider the cone that has a base of radius 4 m and a height of 5 m. Picture the cone lying horizontally with the center of its base at the origin and think of the cone as a solid of revolution.

    i. Write and evaluate a definite integral whose value is the volume of the cone.

    ii. Next, suppose that the cone has uniform density of 800 kg/m^3. What is the mass of the solid cone?

    iii. Now suppose that the cone's density is not uniform, but rather that the cone is most dense at its base. In particular, assume that the density of the cone is uniform across cross sections parallel to its base, but that in each such cross section that is a distance x units from the origin, the density of the cross section is given by the function $\rho(x) = 400 + \frac{200}{1+x^2}$, measured in kg/m^3. Determine and evaluate a definite integral whose value is the mass of this cone of non-uniform density. Do so by first thinking about the mass of a given slice of the cone x units away from the base; remember that in such a slice, the density will be *essentially constant*.

c. Let a thin rod of constant cross-sectional area 1 cm^2 and length 12 cm have its mass be distributed according to the density function $\rho(x) = \frac{1}{25}(x - 15)^2$, measured in g/cm. Find the exact location z at which to cut the bar so that the two pieces will each have identical mass.

Activity 6.3.3. For quantities of equal weight, such as two children on a teeter-totter, the balancing point is found by taking the average of their locations. When the weights of the quantities differ, we use a weighted average of their respective locations to find the balancing point.

a. Suppose that a shelf is 6 feet long, with its left end situated at $x = 0$. If one book of weight 1 lb is placed at $x_1 = 0$, and another book of weight 1 lb is placed at $x_2 = 6$, what is the location of $\overline{x}$, the point at which the shelf would (theoretically) balance on a fulcrum?

b. Now, say that we place four books on the shelf, each weighing 1 lb: at $x_1 = 0$, at $x_2 = 2$, at $x_3 = 4$, and at $x_4 = 6$. Find $\overline{x}$, the balancing point of the shelf.

c. How does $\overline{x}$ change if we change the location of the third book? Say the locations of the 1-lb books are $x_1 = 0$, $x_2 = 2$, $x_3 = 3$, and $x_4 = 6$.

d. Next, suppose that we place four books on the shelf, but of varying weights: at $x_1 = 0$ a 2-lb book, at $x_2 = 2$ a 3-lb book, at $x_3 = 4$ a 1-lb book, and at $x_4 = 6$ a 1-lb book. Use a weighted average of the locations to find $\overline{x}$, the balancing point of the shelf. How does the balancing point in this scenario compare to that found in (b)?

e. What happens if we change the location of one of the books? Say that we keep everything the same in (d), except that $x_3 = 5$. How does $\overline{x}$ change?

f. What happens if we change the weight of one of the books? Say that we keep everything the same in (d), except that the book at $x_3 = 4$ now weighs 2 lbs. How does $\overline{x}$ change?

g. Experiment with a couple of different scenarios of your choosing where you move one of the books to the left, or you decrease the weight of one of the books.

h. Write a couple of sentences to explain how adjusting the location of one of the books or the weight of one of the books affects the location of the balancing point of the shelf. Think carefully here about how your changes should be considered relative to the location of the balancing point $\overline{x}$ of the current scenario.

Activity 6.3.4. Consider a thin bar of length 20 cm whose density is distributed according to the function $\rho(x) = 4 + 0.1x$, where $x = 0$ represents the left end of the bar. Assume that ρ is measured in g/cm and x is measured in cm.

   a. Find the total mass, M, of the bar.

   b. Without doing any calculations, do you expect the center of mass of the bar to be equal to 10, less than 10, or greater than 10? Why?

   c. Compute $\overline{x}$, the exact center of mass of the bar.

   d. What is the average density of the bar?

   e. Now consider a different density function, given by $p(x) = 4e^{0.020732x}$, also for a bar of length 20 cm whose left end is at $x = 0$. Plot both $\rho(x)$ and $p(x)$ on the same axes. Without doing any calculations, which bar do you expect to have the greater center of mass? Why?

   f. Compute the exact center of mass of the bar described in (e) whose density function is $p(x) = 4e^{0.020732x}$. Check the result against the prediction you made in (e).

6.4 Physics Applications: Work, Force, and Pressure

Preview Activity 6.4.1. A bucket is being lifted from the bottom of a 50-foot deep well; its weight (including the water), B, in pounds at a height h feet above the water is given by the function $B(h)$. When the bucket leaves the water, the bucket and water together weigh $B(0) = 20$ pounds, and when the bucket reaches the top of the well, $B(50) = 12$ pounds. Assume that the bucket loses water at a constant rate (as a function of height, h) throughout its journey from the bottom to the top of the well.

a. Find a formula for $B(h)$.

b. Compute the value of the product $B(5)\Delta h$, where $\Delta h = 2$ feet. Include units on your answer. Explain why this product represents the approximate work it took to move the bucket of water from $h = 5$ to $h = 7$.

c. Is the value in (b) an over- or under-estimate of the actual amount of work it took to move the bucket from $h = 5$ to $h = 7$? Why?

d. Compute the value of the product $B(22)\Delta h$, where $\Delta h = 0.25$ feet. Include units on your answer. What is the meaning of the value you found?

e. More generally, what does the quantity $W_{\text{slice}} = B(h)\Delta h$ measure for a given value of h and a small positive value of Δh?

f. Evaluate the definite integral $\int_0^{50} B(h)\,dh$. What is the meaning of the value you find? Why?

Activity 6.4.2. Consider the following situations in which a varying force accomplishes work.

a. Suppose that a heavy rope hangs over the side of a cliff. The rope is 200 feet long and weighs 0.3 pounds per foot; initially the rope is fully extended. How much work is required to haul in the entire length of the rope? (Hint: set up a function $F(h)$ whose value is the weight of the rope remaining over the cliff after h feet have been hauled in.)

b. A leaky bucket is being hauled up from a 100 foot deep well. When lifted from the water, the bucket and water together weigh 40 pounds. As the bucket is being hauled upward at a constant rate, the bucket leaks water at a constant rate so that it is losing weight at a rate of 0.1 pounds per foot. What function $B(h)$ tells the weight of the bucket after the bucket has been lifted h feet? What is the total amount of work accomplished in lifting the bucket to the top of the well?

c. Now suppose that the bucket in (b) does not leak at a constant rate, but rather that its weight at a height h feet above the water is given by $B(h) = 25 + 15e^{-0.05h}$. What is the total work required to lift the bucket 100 feet? What is the average force exerted on the bucket on the interval $h = 0$ to $h = 100$?

d. From physics, *Hooke's Law* for springs states that the amount of force required to hold a spring that is compressed (or extended) to a particular length is proportionate to the distance the spring is compressed (or extended) from its natural length. That is, the force to compress (or extend) a spring x units from its natural length is $F(x) = kx$ for some constant k (which is called the *spring constant*.) For springs, we choose to measure the force in pounds and the distance the spring is compressed in feet. Suppose that a force of 5 pounds extends a particular spring 4 inches (1/3 foot) beyond its natural length.

   i. Use the given fact that $F(1/3) = 5$ to find the spring constant k.

   ii. Find the work done to extend the spring from its natural length to 1 foot beyond its natural length.

   iii. Find the work required to extend the spring from 1 foot beyond its natural length to 1.5 feet beyond its natural length.

Activity 6.4.3. In each of the following problems, determine the total work required to accomplish the described task. In parts (b) and (c), a key step is to find a formula for a function that describes the curve that forms the side boundary of the tank.

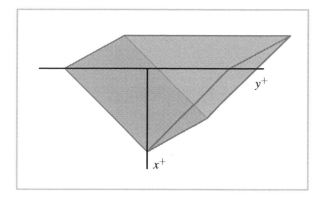

Figure 6.4.5: A trough with triangular ends, as described in Activity 6.4.3, part (c).

a. Consider a vertical cylindrical tank of radius 2 meters and depth 6 meters. Suppose the tank is filled with 4 meters of water of mass density 1000 kg/m^3, and the top 1 meter of water is pumped over the top of the tank.

b. Consider a hemispherical tank with a radius of 10 feet. Suppose that the tank is full to a depth of 7 feet with water of weight density 62.4 pounds/ft^3, and the top 5 feet of water are pumped out of the tank to a tanker truck whose height is 5 feet above the top of the tank.

c. Consider a trough with triangular ends, as pictured in Figure 6.4.5, where the tank is 10 feet long, the top is 5 feet wide, and the tank is 4 feet deep. Say that the trough is full to within 1 foot of the top with water of weight density 62.4 pounds/ft^3, and a pump is used to empty the tank until the water remaining in the tank is 1 foot deep.

Activity 6.4.4. In each of the following problems, determine the total force exerted by water against the surface that is described.

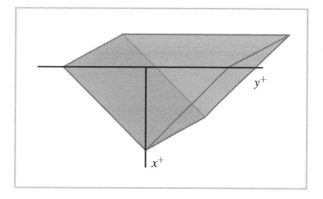

Figure 6.4.8: A trough with triangular ends, as described in Activity 6.4.4, part (c).

a. Consider a rectangular dam that is 100 feet wide and 50 feet tall, and suppose that water presses against the dam all the way to the top.

b. Consider a semicircular dam with a radius of 30 feet. Suppose that the water rises to within 10 feet of the top of the dam.

c. Consider a trough with triangular ends, as pictured in Figure 6.4.8, where the tank is 10 feet long, the top is 5 feet wide, and the tank is 4 feet deep. Say that the trough is full to within 1 foot of the top with water of weight density 62.4 pounds/ft^3. How much force does the water exert against one of the triangular ends?

6.5 Improper Integrals

Preview Activity 6.5.1. A company with a large customer base has a call center that receives thousands of calls a day. After studying the data that represents how long callers wait for assistance, they find that the function $p(t) = 0.25e^{-0.25t}$ models the time customers wait in the following way: the fraction of customers who wait between $t = a$ and $t = b$ minutes is given by

$$\int_a^b p(t)\,dt.$$

Use this information to answer the following questions.

  a. Determine the fraction of callers who wait between 5 and 10 minutes.

  b. Determine the fraction of callers who wait between 10 and 20 minutes.

  c. Next, let's study the fraction who wait up to a certain number of minutes:

    i. What is the fraction of callers who wait between 0 and 5 minutes?
    ii. What is the fraction of callers who wait between 0 and 10 minutes?
    iii. Between 0 and 15 minutes? Between 0 and 20?

  d. Let $F(b)$ represent the fraction of callers who wait between 0 and b minutes. Find a formula for $F(b)$ that involves a definite integral, and then use the First FTC to find a formula for $F(b)$ that does not involve a definite integral.

  e. What is the value of the limit $\lim_{b\to\infty} F(b)$? What is its meaning in the context of the problem?

Activity 6.5.2. In this activity we explore the improper integrals $\int_1^\infty \frac{1}{x}\, dx$ and $\int_1^\infty \frac{1}{x^{3/2}}\, dx$.

a. First we investigate $\int_1^\infty \frac{1}{x}\, dx$.

    i. Use the First FTC to determine the exact values of $\int_1^{10} \frac{1}{x}\, dx$, $\int_1^{1000} \frac{1}{x}\, dx$, and $\int_1^{100000} \frac{1}{x}\, dx$. Then, use your computational device to compute a decimal approximation of each result.

    ii. Use the First FTC to evaluate the definite integral $\int_1^b \frac{1}{x}\, dx$ (which results in an expression that depends on b).

    iii. Now, use your work from (ii.) to evaluate the limit given by

$$\lim_{b \to \infty} \int_1^b \frac{1}{x}\, dx.$$

b. Next, we investigate $\int_1^\infty \frac{1}{x^{3/2}}\, dx$.

    i. Use the First FTC to determine the exact values of $\int_1^{10} \frac{1}{x^{3/2}}\, dx$, $\int_1^{1000} \frac{1}{x^{3/2}}\, dx$, and $\int_1^{100000} \frac{1}{x^{3/2}}\, dx$. Then, use your calculator to compute a decimal approximation of each result.

    ii. Use the First FTC to evaluate the definite integral $\int_1^b \frac{1}{x^{3/2}}\, dx$ (which results in an expression that depends on b).

    iii. Now, use your work from (ii.) to evaluate the limit given by

$$\lim_{b \to \infty} \int_1^b \frac{1}{x^{3/2}}\, dx.$$

c. Plot the functions $y = \frac{1}{x}$ and $y = \frac{1}{x^{3/2}}$ on the same coordinate axes for the values $x = 0 \ldots 10$. How would you compare their behavior as x increases without bound? What is similar? What is different?

d. How would you characterize the value of $\int_1^\infty \frac{1}{x}\, dx$? of $\int_1^\infty \frac{1}{x^{3/2}}\, dx$? What does this tell us about the respective areas bounded by these two curves for $x \geq 1$?

Activity 6.5.3. Determine whether each of the following improper integrals converges or diverges. For each integral that converges, find its exact value.

a. $\int_1^\infty \frac{1}{x^2}\,dx$

b. $\int_0^\infty e^{-x/4}\,dx$

c. $\int_2^\infty \frac{9}{(x+5)^{2/3}}\,dx$

d. $\int_4^\infty \frac{3}{(x+2)^{5/4}}\,dx$

e. $\int_0^\infty xe^{-x/4}\,dx$

f. $\int_1^\infty \frac{1}{x^p}\,dx$, where p is a positive real number

Activity 6.5.4. For each of the following definite integrals, decide whether the integral is improper or not. If the integral is proper, evaluate it using the First FTC. If the integral is improper, determine whether or not the integral converges or diverges; if the integral converges, find its exact value.

a. $\int_0^1 \frac{1}{x^{1/3}}\, dx$

b. $\int_0^2 e^{-x}\, dx$

c. $\int_1^4 \frac{1}{\sqrt{4-x}}\, dx$

d. $\int_{-2}^2 \frac{1}{x^2}\, dx$

e. $\int_0^{\pi/2} \tan(x)\, dx$

f. $\int_0^1 \frac{1}{\sqrt{1-x^2}}\, dx$

CHAPTER 7

Differential Equations

7.1 An Introduction to Differential Equations

Preview Activity 7.1.1. The position of a moving object is given by the function $s(t)$, where s is measured in feet and t in seconds. We determine that the velocity is $v(t) = 4t + 1$ feet per second.

    a. How much does the position change over the time interval $[0, 4]$?

    b. Does this give you enough information to determine $s(4)$, the position at time $t = 4$? If so, what is $s(4)$? If not, what additional information would you need to know to determine $s(4)$?

    c. Suppose you are told that the object's initial position $s(0) = 7$. Determine $s(2)$, the object's position 2 seconds later.

    d. If you are told instead that the object's initial position is $s(0) = 3$, what is $s(2)$?

    e. If we only know the velocity $v(t) = 4t + 1$, is it possible that the object's position at all times is $s(t) = 2t^2 + t - 4$? Explain how you know.

    f. Are there other possibilities for $s(t)$? If so, what are they?

    g. If, in addition to knowing the velocity function is $v(t) = 4t + 1$, we know the initial position $s(0)$, how many possibilities are there for $s(t)$?

Activity 7.1.2. Express the following statements as differential equations. In each case, you will need to introduce notation to describe the important quantities in the statement so be sure to clearly state what your notation means.

a. The population of a town grows continuously at an annual rate of 1.25%.

b. A radioactive sample loses mass at a rate of 5.6% of its mass every day.

c. You have a bank account that continuously earns 4% interest every year. At the same time, you withdraw money continually from the account at the rate of $1000 per year.

d. A cup of hot chocolate is sitting in a 70° room. The temperature of the hot chocolate cools continuously by 10% of the difference between the hot chocolate's temperature and the room temperature every minute.

e. A can of cold soda is sitting in a 70° room. The temperature of the soda warms continuously at the rate of 10% of the difference between the soda's temperature and the room's temperature every minute.

Activity 7.1.3. Shown below are two graphs depicting the velocity of falling objects. On the left is the velocity of a skydiver, while on the right is the velocity of a meteorite entering the Earth's atmosphere.

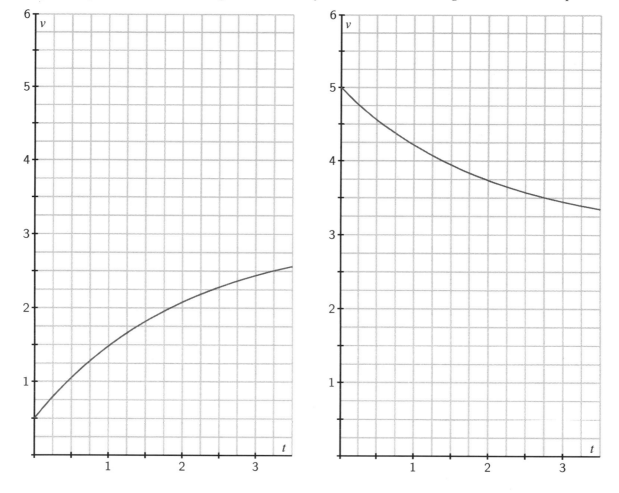

Figure 7.1.1: A skydiver's velocity. Figure 7.1.2: A meteorite's velocity.

a. Begin with the skydiver's velocity and use the given graph to measure the rate of change dv/dt when the velocity is $v = 0.5, 1.0, 1.5, 2.0$, and 2.5. Plot your values on the graph below. You will want to think carefully about this: you are plotting the derivative dv/dt as a function of *velocity*.

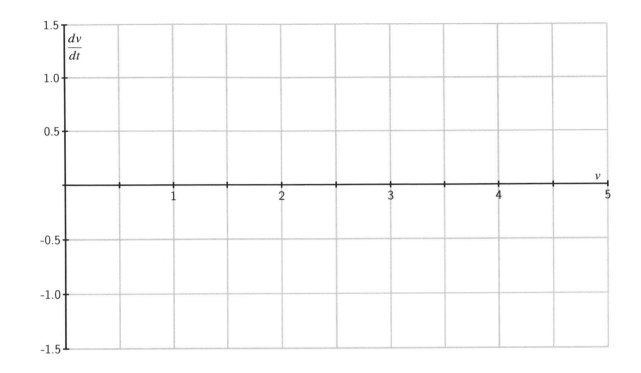

b. Now do the same thing with the meteorite's velocity: use the given graph to measure the rate of change dv/dt when the velocity is $v = 3.5, 4.0, 4.5,$ and 5.0. Plot your values on the graph above.

c. You should find that all your points lie on a line. Write the equation of this line being careful to use proper notation for the quantities on the horizontal and vertical axes.

d. The relationship you just found is a differential equation. Write a complete sentence that explains its meaning.

e. By looking at the differential equation, determine the values of the velocity for which the velocity increases.

f. By looking at the differential equation, determine the values of the velocity for which the velocity decreases.

g. By looking at the differential equation, determine the values of the velocity for which the velocity remains constant.

Activity 7.1.4. Consider the differential equation

$$\frac{dv}{dt} = 1.5 - 0.5v.$$

Which of the following functions are solutions of this differential equation?

a. $v(t) = 1.5t - 0.25t^2$.

b. $v(t) = 3 + 2e^{-0.5t}$.

c. $v(t) = 3$.

d. $v(t) = 3 + Ce^{-0.5t}$ where C is any constant.

7.2 Qualitative behavior of solutions to DEs

Preview Activity 7.2.1. Let's consider the initial value problem

$$\frac{dy}{dt} = t - 2, \quad y(0) = 1.$$

a. Use the differential equation to find the slope of the tangent line to the solution $y(t)$ at $t = 0$. Then use the initial value to find the equation of the tangent line at $t = 0$. Sketch this tangent line over the interval $-0.25 \le t \le 0.25$ on the axes provided in Figure 7.2.1.

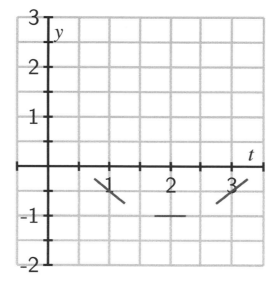

Figure 7.2.1: Grid for plotting partial tangent lines.

b. Also shown in Figure 7.2.1 are the tangent lines to the solution $y(t)$ at the points $t = 1, 2$, and 3 (we will see how to find these later). Use the graph to measure the slope of each tangent line and verify that each agrees with the value specified by the differential equation.

c. Using these tangent lines as a guide, sketch a graph of the solution $y(t)$ over the interval $0 \le t \le 3$ so that the lines are tangent to the graph of $y(t)$.

d. Use the Fundamental Theorem of Calculus to find $y(t)$, the solution to this initial value problem.

e. Graph the solution you found in (d) on the axes provided, and compare it to the sketch you made using the tangent lines.

Activity 7.2.2. Consider the autonomous differential equation

$$\frac{dy}{dt} = -\frac{1}{2}(y-4).$$

a. Make a plot of $\frac{dy}{dt}$ versus y on the axes provided in Figure 7.2.8. Looking at the graph, for what values of y does y increase and for what values of y does y decrease?

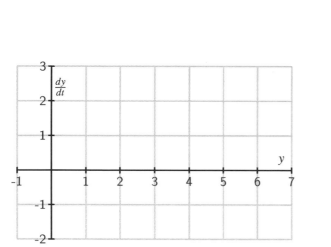

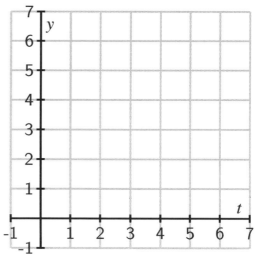

Figure 7.2.8: Axes for plotting $\frac{dy}{dt}$ versus y.

Figure 7.2.9: Axes for plotting the slope field for $\frac{dy}{dt} = -\frac{1}{2}(y-4)$.

b. Next, sketch the slope field for this differential equation on the axes provided in Figure 7.2.9.

c. Use your work in (b) to sketch (on the same axes in Figure 7.2.9.) solutions that satisfy $y(0) = 0$, $y(0) = 2$, $y(0) = 4$ and $y(0) = 6$.

d. Verify that $y(t) = 4 + 2e^{-t/2}$ is a solution to the given differential equation with the initial value $y(0) = 6$. Compare its graph to the one you sketched in (c).

e. What is special about the solution where $y(0) = 4$?

Activity 7.2.3. Consider the autonomous differential equation

$$\frac{dy}{dt} = -\frac{1}{2}y(y-4).$$

a. Make a plot of $\frac{dy}{dt}$ versus y on the axes provided in Figure 7.2.10. Looking at the graph, for what values of y does y increase and for what values of y does y decrease?

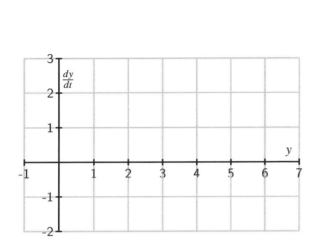

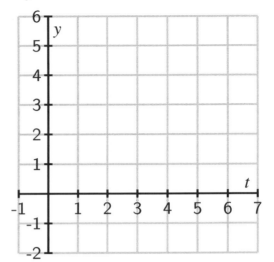

Figure 7.2.10: Axes for plotting dy/dt vs y for $\frac{dy}{dt} = -\frac{1}{2}y(y-4)$.

Figure 7.2.11: Axes for plotting the slope field for $\frac{dy}{dt} = -\frac{1}{2}y(y-4)$.

b. Identify any equilibrium solutions of the given differential equation.

c. Now sketch the slope field for the given differential equation on the axes provided in Figure 7.2.11.

d. Sketch the solutions to the given differential equation that correspond to initial values $y(0) = -1, 0, 1, \ldots, 5$.

e. An equilibrium solution $\overline{y}$ is called *stable* if nearby solutions converge to $\overline{y}$. This means that if the initial condition varies slightly from $\overline{y}$, then $\lim_{t \to \infty} y(t) = \overline{y}$. Conversely, an equilibrium solution $\overline{y}$ is called *unstable* if nearby solutions are pushed away from $\overline{y}$. Using your work above, classify the equilibrium solutions you found in (b) as either stable or unstable.

f. Suppose that $y(t)$ describes the population of a species of living organisms and that the initial value $y(0)$ is positive. What can you say about the eventual fate of this population?

g. Now consider a general autonomous differential equation of the form $dy/dt = f(y)$. Remember that an equilibrium solution $\overline{y}$ satisfies $f(\overline{y}) = 0$. If we graph $dy/dt = f(y)$ as a function of y, for which of the differential equations represented in Figure 7.2.12 and Figure 7.2.13 is $\overline{y}$ a stable equilibrium and for which is $\overline{y}$ unstable? Why?

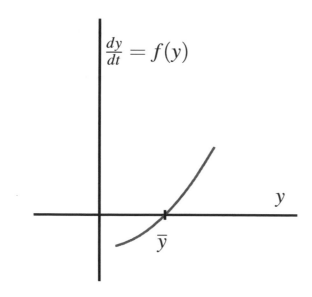

Figure 7.2.12: Plot of $\frac{dy}{dt}$ as a function of y.

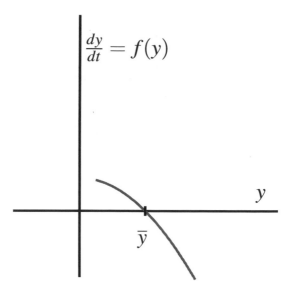

Figure 7.2.13: Plot of $\frac{dy}{dt}$ as a different function of y.

7.3 Euler's method

Preview Activity 7.3.1. Consider the initial value problem

$$\frac{dy}{dt} = \frac{1}{2}(y + 1), \ y(0) = 0.$$

a. Use the differential equation to find the slope of the tangent line to the solution $y(t)$ at $t = 0$. Then use the given initial value to find the equation of the tangent line at $t = 0$.

b. Sketch the tangent line on the axes provided in Figure 7.3.1 on the interval $0 \le t \le 2$ and use it to approximate $y(2)$, the value of the solution at $t = 2$.

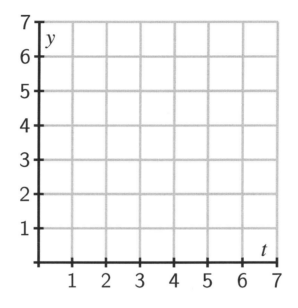

Figure 7.3.1: Grid for plotting the tangent line.

c. Assuming that your approximation for $y(2)$ is the actual value of $y(2)$, use the differential equation to find the slope of the tangent line to $y(t)$ at $t = 2$. Then, write the equation of the tangent line at $t = 2$.

d. Add a sketch of this tangent line on the interval $2 \le t \le 4$ to your plot Figure 7.3.1; use this new tangent line to approximate $y(4)$, the value of the solution at $t = 4$.

e. Repeat the same step to find an approximation for $y(6)$.

Activity 7.3.2. Consider the initial value problem

$$\frac{dy}{dt} = 2t - 1, \; y(0) = 0$$

a. Use Euler's method with $\Delta t = 0.2$ to approximate the solution at $t_i = 0.2, 0.4, 0.6, 0.8,$ and 1.0. Record your work in the following table, and sketch the points (t_i, y_i) on the axes provided.

t_i	y_i	dy/dt	Δy
0.0000	0.0000		
0.2000			
0.4000			
0.6000			
0.8000			
1.0000			

Table 7.3.8: Table for recording results of Euler's method.

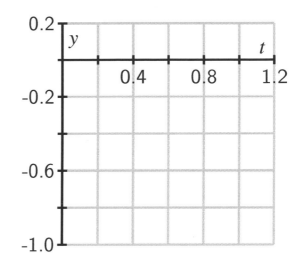

Figure 7.3.9: Grid for plotting points generated by Euler's method.

b. Find the exact solution to the original initial value problem and use this function to find the error in your approximation at each one of the points t_i.

c. Explain why the value y_5 generated by Euler's method for this initial value problem produces the same value as a left Riemann sum for the definite integral $\int_0^1 (2t - 1)\, dt$.

d. How would your computations differ if the initial value was $y(0) = 1$? What does this mean about different solutions to this differential equation?

Activity 7.3.3. Consider the differential equation $\frac{dy}{dt} = 6y - y^2$.

  a. Sketch the slope field for this differential equation on the axes provided in Figure 7.3.10.

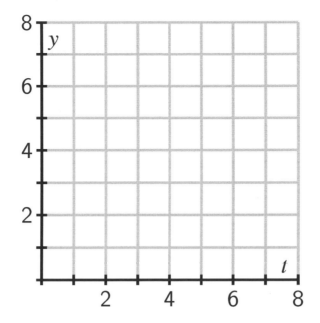

Figure 7.3.10: Grid for plotting the slope field of the given differential equation.

  b. Identify any equilibrium solutions and determine whether they are stable or unstable.

  c. What is the long-term behavior of the solution that satisfies the initial value $y(0) = 1$?

  d. Using the initial value $y(0) = 1$, use Euler's method with $\Delta t = 0.2$ to approximate the solution at $t_i = 0.2, 0.4, 0.6, 0.8$, and 1.0. Record your results in Table 7.3.11 and sketch the corresponding points (t_i, y_i) on the axes provided in Figure 7.3.12. Note the different horizontal scale on the axes in Figure 7.3.12 compared to Figure 7.3.10.

t_i	y_i	dy/dt	Δy
0.0	1.0000		
0.2			
0.4			
0.6			
0.8			
1.0			

Table 7.3.11: Table for recording results of Euler's method with $\Delta t = 0.2$.

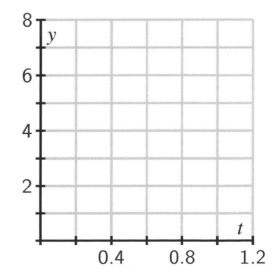

Figure 7.3.12: Axes for plotting the results of Euler's method.

  e. What happens if we apply Euler's method to approximate the solution with $y(0) = 6$?

7.4 Separable differential equations

Preview Activity 7.4.1. In this preview activity, we explore whether certain differential equations are separable or not, and then revisit some key ideas from earlier work in integral calculus.

   a. Which of the following differential equations are separable? If the equation is separable, write the equation in the revised form $g(y)\frac{dy}{dt} = h(t)$.

   i. $\dfrac{dy}{dt} = -3y.$

   ii. $\dfrac{dy}{dt} = ty - y.$

   iii. $\dfrac{dy}{dt} = t + 1.$

   iv. $\dfrac{dy}{dt} = t^2 - y^2.$

   b. Explain why any autonomous differential equation is guaranteed to be separable.

   c. Why do we include the term "+C" in the expression

$$\int x \, dx = \frac{x^2}{2} + C?$$

   d. Suppose we know that a certain function f satisfies the equation

$$\int f'(x) \, dx = \int x \, dx.$$

What can you conclude about f?

Activity 7.4.2. Suppose that the population of a town is growing continuously at an annual rate of 3% per year.

    a. Let $P(t)$ be the population of the town in year t. Write a differential equation that describes the annual growth rate.

    b. Find the solutions of this differential equation.

    c. If you know that the town's population in year 0 is 10,000, find the population $P(t)$.

    d. How long does it take for the population to double? This time is called the *doubling time*.

    e. Working more generally, find the doubling time if the annual growth rate is k times the population.

Activity 7.4.3. Suppose that a cup of coffee is initially at a temperature of 105° F and is placed in a 75° F room. Newton's law of cooling says that

$$\frac{dT}{dt} = -k(T - 75),$$

where k is a constant of proportionality.

a. Suppose you measure that the coffee is cooling at one degree per minute at the time the coffee is brought into the room. Use the differential equation to determine the value of the constant k.

b. Find all the solutions of this differential equation.

c. What happens to all the solutions as $t \to \infty$? Explain how this agrees with your intuition.

d. What is the temperature of the cup of coffee after 20 minutes?

e. How long does it take for the coffee to cool to 80°?

Activity 7.4.4. Solve each of the following differential equations or initial value problems.

a. $\frac{dy}{dt} - (2-t)y = 2-t$

b. $\frac{1}{t}\frac{dy}{dt} = e^{t^2-2y}$

c. $y' = 2y+2, \quad y(0) = 2$

d. $y' = 2y^2, \quad y(-1) = 2$

e. $\frac{dy}{dt} = \frac{-2ty}{t^2+1}, \quad y(0) = 4$

7.5 Modeling with differential equations

Preview Activity 7.5.1. Any time that the rate of change of a quantity is related to the amount of a quantity, a differential equation naturally arises. In the following two problems, we see two such scenarios; for each, we want to develop a differential equation whose solution is the quantity of interest.

a. Suppose you have a bank account in which money grows at an annual rate of 3%.

    i. If you have $10,000 in the account, at what rate is your money growing?

    ii. Suppose that you are also withdrawing money from the account at $1,000 per year. What is the rate of change in the amount of money in the account? What are the units on this rate of change?

b. Suppose that a water tank holds 100 gallons and that a salty solution, which contains 20 grams of salt in every gallon, enters the tank at 2 gallons per minute.

    i. How much salt enters the tank each minute?

    ii. Suppose that initially there are 300 grams of salt in the tank. How much salt is in each gallon at this point in time?

    iii. Finally, suppose that evenly mixed solution is pumped out of the tank at the rate of 2 gallons per minute. How much salt leaves the tank each minute?

    iv. What is the total rate of change in the amount of salt in the tank?

Activity 7.5.2. Suppose you have a bank account that grows by 5% every year. Let $A(t)$ be the amount of money in the account in year t.

  a. What is the rate of change of A with respect to t?

  b. Suppose that you are also withdrawing $10,000 per year. Write a differential equation that expresses the total rate of change of A.

  c. Sketch a slope field for this differential equation, find any equilibrium solutions, and identify them as either stable or unstable. Write a sentence or two that describes the significance of the stability of the equilibrium solution.

  d. Suppose that you initially deposit $100,000 into the account. How long does it take for you to deplete the account?

  e. What is the smallest amount of money you would need to have in the account to guarantee that you never deplete the money in the account?

  f. If your initial deposit is $300,000, how much could you withdraw every year without depleting the account?

Activity 7.5.3. A dose of morphine is absorbed from the bloodstream of a patient at a rate proportional to the amount in the bloodstream.

    a. Write a differential equation for $M(t)$, the amount of morphine in the patient's bloodstream, using k as the constant proportionality.

    b. Assuming that the initial dose of morphine is M_0, solve the initial value problem to find $M(t)$. Use the fact that the half-life for the absorption of morphine is two hours to find the constant k.

    c. Suppose that a patient is given morphine intravenously at the rate of 3 milligrams per hour. Write a differential equation that combines the intravenous administration of morphine with the body's natural absorption.

    d. Find any equilibrium solutions and determine their stability.

    e. Assuming that there is initially no morphine in the patient's bloodstream, solve the initial value problem to determine $M(t)$. What happens to $M(t)$ after a very long time?

    f. To what rate should a doctor reduce the intravenous rate so that there is eventually 7 milligrams of morphine in the patient's bloodstream?

7.6 Population Growth and the Logistic Equation

Preview Activity 7.6.1. Recall that one model for population growth states that a population grows at a rate proportional to its size.

a. We begin with the differential equation

$$\frac{dP}{dt} = \frac{1}{2}P.$$

Sketch a slope field below as well as a few typical solutions on the axes provided.

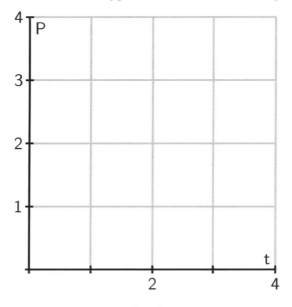

b. Find all equilibrium solutions of the equation $\frac{dP}{dt} = \frac{1}{2}P$ and classify them as stable or unstable.

c. If $P(0)$ is positive, describe the long-term behavior of the solution to $\frac{dP}{dt} = \frac{1}{2}P$.

d. Let's now consider a modified differential equation given by

$$\frac{dP}{dt} = \frac{1}{2}P(3 - P).$$

As before, sketch a slope field as well as a few typical solutions on the following axes provided.

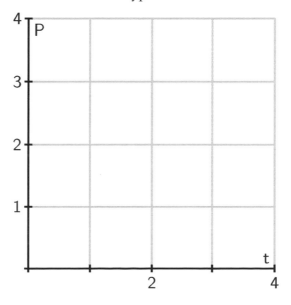

e. Find any equilibrium solutions and classify them as stable or unstable.

f. If $P(0)$ is positive, describe the long-term behavior of the solution.

Activity 7.6.2. Our first model will be based on the following assumption:

> *The rate of change of the population is proportional to the population.*

On the face of it, this seems pretty reasonable. When there is a relatively small number of people, there will be fewer births and deaths so the rate of change will be small. When there is a larger number of people, there will be more births and deaths so we expect a larger rate of change.

If $P(t)$ is the population t years after the year 2000, we may express this assumption as

$$\frac{dP}{dt} = kP$$

where k is a constant of proportionality.

a. Use the data in the table to estimate the derivative $P'(0)$ using a central difference. Assume that $t = 0$ corresponds to the year 2000.

b. What is the population $P(0)$?

c. Use your results from (a) and (b) to estimate the constant of proportionality k in the differential equation.

d. Now that we know the value of k, we have the initial value problem

$$\frac{dP}{dt} = kP, \; P(0) = 6.084.$$

Find the solution to this initial value problem.

e. What does your solution predict for the population in the year 2010? Is this close to the actual population given in the table?

f. When does your solution predict that the population will reach 12 billion?

g. What does your solution predict for the population in the year 2500?

h. Do you think this is a reasonable model for the earth's population? Why or why not? Explain your thinking using a couple of complete sentences.

Activity 7.6.3. Consider the logistic equation

$$\frac{dP}{dt} = kP(N - P)$$

with the graph of $\frac{dP}{dt}$ vs. P shown in Figure 7.6.6.

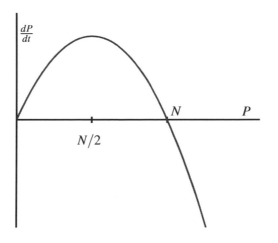

Figure 7.6.6: Plot of $\frac{dP}{dt}$ vs. P.

a. At what value of P is the rate of change greatest?

b. Consider the model for the earth's population that we created. At what value of P is the rate of change greatest? How does that compare to the population in recent years?

c. According to the model we developed, what will the population be in the year 2100?

d. According to the model we developed, when will the population reach 9 billion?

e. Now consider the general solution to the general logistic initial value problem that we found, given by

$$P(t) = \frac{N}{\left(\frac{N-P_0}{P_0}\right) e^{-kNt} + 1}.$$

Verify algebraically that $P(0) = P_0$ and that $\lim_{t\to\infty} P(t) = N$.

CHAPTER 8

Sequences and Series

8.1 Sequences

Preview Activity 8.1.1. Suppose you receive $5000 through an inheritance. You decide to invest this money into a fund that pays 8% annually, compounded monthly. That means that each month your investment earns $\frac{0.08}{12} \cdot P$ additional dollars, where P is your principal balance at the start of the month. So in the first month your investment earns

$$5000 \left(\frac{0.08}{12} \right)$$

or $33.33. If you reinvest this money, you will then have $5033.33 in your account at the end of the first month. From this point on, assume that you reinvest all of the interest you earn.

   a. How much interest will you earn in the second month? How much money will you have in your account at the end of the second month?

   b. Complete Table 8.1.2 to determine the interest earned and total amount of money in this investment each month for one year.

Month	Interest earned	Total amount of money in the account
0	$0.00	$5000.00
1	$33.33	$5033.33
2		
3		
4		
5		
6		
7		
8		
9		
10		
11		
12		

Table 8.1.2: Interest

c. As we will see later, the amount of money P_n in the account after month n is given by

$$P_n = 5000 \left(1 + \frac{0.08}{12}\right)^n.$$

Use this formula to check your calculations in Table 8.1.2. Then find the amount of money in the account after 5 years.

d. How many years will it be before the account has doubled in value to $10000?

Activity 8.1.2.

a. Let s_n be the nth term in the sequence $1, 2, 3, \ldots$. Find a formula for s_n and use appropriate technological tools to draw a graph of entries in this sequence by plotting points of the form (n, s_n) for some values of n. Most graphing calculators can plot sequences; directions follow for the TI-84.

   - In the MODEmenu, highlight SEQin the FUNCline and press ENTER.

   - In the Y=menu, you will now see lines to enter sequences. Enter a value for nMin (where the sequence starts), a function for u(n) (the nth term in the sequence), and the value of u(nMin).

   - Set your window coordinates (this involves choosing limits for n as well as the window coordinates XMin, XMax, YMin, and YMax.

   - The GRAPHkey will draw a plot of your sequence.

   Using your knowledge of limits of continuous functions as $x \to \infty$, decide if this sequence $\{s_n\}$ has a limit as $n \to \infty$. Explain your reasoning.

b. Let s_n be the nth term in the sequence $1, \frac{1}{2}, \frac{1}{3}, \ldots$. Find a formula for s_n. Draw a graph of some points in this sequence. Using your knowledge of limits of continuous functions as $x \to \infty$, decide if this sequence $\{s_n\}$ has a limit as $n \to \infty$. Explain your reasoning.

c. Let s_n be the nth term in the sequence $2, \frac{3}{2}, \frac{4}{3}, \frac{5}{4}, \ldots$. Find a formula for s_n. Using your knowledge of limits of continuous functions as $x \to \infty$, decide if this sequence $\{s_n\}$ has a limit as $n \to \infty$. Explain your reasoning.

Activity 8.1.3.

a. Recall our earlier work with limits involving infinity in ((((Unresolved xref, reference "sec-2-8-LHR"; check spelling or use "provisional" attribute)))Section . State clearly what it means for a continuous function f to have a limit L as $x \to \infty$.

b. Given that an infinite sequence of real numbers is a function from the integers to the real numbers, apply the idea from part (a) to explain what you think it means for a sequence $\{s_n\}$ to have a limit as $n \to \infty$.

c. Based on your response to the part (b), decide if the sequence $\left\{\frac{1+n}{2+n}\right\}$ has a limit as $n \to \infty$. If so, what is the limit? If not, why not?

Activity 8.1.4. Use graphical and/or algebraic methods to determine whether each of the following sequences converges or diverges.

a. $\left\{\frac{1+2n}{3n-2}\right\}$

b. $\left\{\frac{5+3^n}{10+2^n}\right\}$

c. $\left\{\frac{10^n}{n!}\right\}$ (where ! is the *factorial* symbol and $n! = n(n-1)(n-2)\cdots(2)(1)$ for any positive integer n (as convention we define 0! to be 1)).

8.2 Geometric Series

Preview Activity 8.2.1. Warfarin is an anticoagulant that prevents blood clotting; often it is prescribed to stroke victims in order to help ensure blood flow. The level of warfarin has to reach a certain concentration in the blood in order to be effective.

Suppose warfarin is taken by a particular patient in a 5 mg dose each day. The drug is absorbed by the body and some is excreted from the system between doses. Assume that at the end of a 24 hour period, 8% of the drug remains in the body. Let $Q(n)$ be the amount (in mg) of warfarin in the body before the $(n + 1)$st dose of the drug is administered.

    a. Explain why $Q(1) = 5 \times 0.08$ mg.

    b. Explain why $Q(2) = (5 + Q(1)) \times 0.08$ mg. Then show that

$$Q(2) = (5 \times 0.08)(1 + 0.08) \text{ mg.}$$

    c. Explain why $Q(3) = (5 + Q(2)) \times 0.08$ mg. Then show that

$$Q(3) = (5 \times 0.08)\left(1 + 0.08 + 0.08^2\right) \text{ mg.}$$

    d. Explain why $Q(4) = (5 + Q(3)) \times 0.08$ mg. Then show that

$$Q(4) = (5 \times 0.08)\left(1 + 0.08 + 0.08^2 + 0.08^3\right) \text{ mg.}$$

    e. There is a pattern that you should see emerging. Use this pattern to find a formula for $Q(n)$, where n is an arbitrary positive integer.

    f. Complete Table 8.2.1 with values of $Q(n)$ for the provided n-values (reporting $Q(n)$ to 10 decimal places). What appears to be happening to the sequence $Q(n)$ as n increases?

n	1	2	3	4	5	6	7	8	9	10
$Q(n)$	0.40									

Table 8.2.1: Values of $Q(n)$ for selected values of n

Activity 8.2.2. Let a and r be real numbers (with $r \neq 1$) and let

$$S_n = a + ar + ar^2 + \cdots + ar^{n-1}.$$

In this activity we will find a shortcut formula for S_n that does not involve a sum of n terms.

    a. Multiply S_n by r. What does the resulting sum look like?

    b. Subtract rS_n from S_n and explain why

$$S_n - rS_n = a - ar^n. \tag{8.2.1}$$

    c. Solve equation (8.2.2) for S_n to find a simple formula for S_n that does not involve adding n terms.

Activity 8.2.3. Let $r \neq 1$ and a be real numbers and let

$$S = a + ar + ar^2 + \cdots ar^{n-1} + \cdots$$

be an infinite geometric series. For each positive integer n, let

$$S_n = a + ar + ar^2 + \cdots + ar^{n-1}.$$

Recall that

$$S_n = a\frac{1 - r^n}{1 - r}.$$

a. What should we allow n to approach in order to have S_n approach S?

b. What is the value of $\lim_{n \to \infty} r^n$ for $|r| > 1$? for $|r| < 1$? Explain.

c. If $|r| < 1$, use the formula for S_n and your observations in (a) and (b) to explain why S is finite and find a resulting formula for S.

Activity 8.2.4. The formulas we have derived for an infinite geometric series and its partial sum have assumed we begin indexing the sums at $n = 0$. If instead we have a sum that does not begin at $n = 0$, we can factor out common terms and use the established formulas. This process is illustrated in the examples in this activity.

a. Consider the sum

$$\sum_{k=1}^{\infty} (2)\left(\frac{1}{3}\right)^k = (2)\left(\frac{1}{3}\right) + (2)\left(\frac{1}{3}\right)^2 + (2)\left(\frac{1}{3}\right)^3 + \cdots.$$

Remove the common factor of $(2)\left(\frac{1}{3}\right)$ from each term and hence find the sum of the series.

b. Next let a and r be real numbers with $-1 < r < 1$. Consider the sum

$$\sum_{k=3}^{\infty} ar^k = ar^3 + ar^4 + ar^5 + \cdots.$$

Remove the common factor of ar^3 from each term and find the sum of the series.

c. Finally, we consider the most general case. Let a and r be real numbers with $-1 < r < 1$, let n be a positive integer, and consider the sum

$$\sum_{k=n}^{\infty} ar^k = ar^n + ar^{n+1} + ar^{n+2} + \cdots.$$

Remove the common factor of ar^n from each term to find the sum of the series.

8.3 Series of Real Numbers

Preview Activity 8.3.1. Have you ever wondered how your calculator can produce a numeric approximation for complicated numbers like e, π or $\ln(2)$? After all, the only operations a calculator can really perform are addition, subtraction, multiplication, and division, the operations that make up polynomials. This activity provides the first steps in understanding how this process works. Throughout the activity, let $f(x) = e^x$.

    a. Find the tangent line to f at $x = 0$ and use this linearization to approximate e. That is, find a formula $L(x)$ for the tangent line, and compute $L(1)$, since $L(1) \approx f(1) = e$.

    b. The linearization of e^x does not provide a good approximation to e since 1 is not very close to 0. To obtain a better approximation, we alter our approach a bit. Instead of using a straight line to approximate e, we put an appropriate bend in our estimating function to make it better fit the graph of e^x for x close to 0. With the linearization, we had both $f(x)$ and $f'(x)$ share the same value as the linearization at $x = 0$. We will now use a quadratic approximation $P_2(x)$ to $f(x) = e^x$ centered at $x = 0$ which has the property that $P_2(0) = f(0)$, $P_2'(0) = f'(0)$, and $P_2''(0) = f''(0)$.

        i) Let $P_2(x) = 1 + x + \frac{x^2}{2}$. Show that $P_2(0) = f(0)$, $P_2'(0) = f'(0)$, and $P_2''(0) = f''(0)$. Then, use $P_2(x)$ to approximate e by observing that $P_2(1) \approx f(1)$.

        ii) We can continue approximating e with polynomials of larger degree whose higher derivatives agree with those of f at 0. This turns out to make the polynomials fit the graph of f better for more values of x around 0. For example, let $P_3(x) = 1 + x + \frac{x^2}{2} + \frac{x^3}{6}$. Show that $P_3(0) = f(0)$, $P_3'(0) = f'(0)$, $P_3''(0) = f''(0)$, and $P_3'''(0) = f'''(0)$. Use $P_3(x)$ to approximate e in a way similar to how you did so with $P_2(x)$ above.

Activity 8.3.2. Consider the series

$$\sum_{k=1}^{\infty} \frac{1}{k^2}.$$

While it is physically impossible to add an infinite collection of numbers, we can, of course, add any finite collection of them. In what follows, we investigate how understanding how to find the nth partial sum (that is, the sum of the first n terms) enables us to make sense of the infinite sum.

a. Sum the first two numbers in this series. That is, find a numeric value for

$$\sum_{k=1}^{2} \frac{1}{k^2}$$

b. Next, add the first three numbers in the series.

c. Continue adding terms in this series to complete the list below. Carry each sum to at least 8 decimal places.

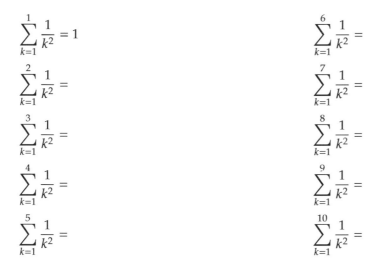

$$\sum_{k=1}^{1} \frac{1}{k^2} = 1 \qquad\qquad \sum_{k=1}^{6} \frac{1}{k^2} =$$

$$\sum_{k=1}^{2} \frac{1}{k^2} = \qquad\qquad \sum_{k=1}^{7} \frac{1}{k^2} =$$

$$\sum_{k=1}^{3} \frac{1}{k^2} = \qquad\qquad \sum_{k=1}^{8} \frac{1}{k^2} =$$

$$\sum_{k=1}^{4} \frac{1}{k^2} = \qquad\qquad \sum_{k=1}^{9} \frac{1}{k^2} =$$

$$\sum_{k=1}^{5} \frac{1}{k^2} = \qquad\qquad \sum_{k=1}^{10} \frac{1}{k^2} =$$

d. The sums in the table in part c form a sequence whose nth term is $S_n = \sum_{k=1}^{n} \frac{1}{k^2}$. Based on your calculations in the table, do you think the sequence $\{S_n\}$ converges or diverges? Explain. How do you think this sequence $\{S_n\}$ is related to the series $\sum_{k=1}^{\infty} \frac{1}{k^2}$?

Activity 8.3.3. If the *series* $\sum a_k$ converges, then an important result necessarily follows regarding the *sequence* $\{a_n\}$. This activity explores this result.

Assume that the series $\sum_{k=1}^{\infty} a_k$ converges and has sum equal to L.

a. What is the nth partial sum S_n of the series $\sum_{k=1}^{\infty} a_k$?

b. What is the $(n-1)$st partial sum S_{n-1} of the series $\sum_{k=1}^{\infty} a_k$?

c. What is the difference between the nth partial sum and the $(n-1)$st partial sum of the series $\sum_{k=1}^{\infty} a_k$?

d. Since we are assuming that $\sum_{k=1}^{\infty} a_k = L$, what does that tell us about $\lim_{n\to\infty} S_n$? Why? What does that tell us about $\lim_{n\to\infty} S_{n-1}$? Why?

e. Combine the results of the previous two parts of this activity to determine $\lim_{n\to\infty} a_n = \lim_{n\to\infty}(S_n - S_{n-1})$.

Activity 8.3.4. Determine if the Divergence Test applies to the following series. If the test does not apply, explain why. If the test does apply, what does it tell us about the series?

a. $\sum \frac{k}{k+1}$
b. $\sum (-1)^k$
c. $\sum \frac{1}{k}$

Activity 8.3.5. Consider the harmonic series $\sum_{k=1}^{\infty} \frac{1}{k}$. Recall that the harmonic series will converge provided that its sequence of partial sums converges. The nth partial sum S_n of the series $\sum_{k=1}^{\infty} \frac{1}{k}$ is

$$S_n = \sum_{k=1}^{n} \frac{1}{k}$$
$$= 1 + \frac{1}{2} + \frac{1}{3} + \cdots + \frac{1}{n}$$
$$= 1(1) + (1)\left(\frac{1}{2}\right) + (1)\left(\frac{1}{3}\right) + \cdots + (1)\left(\frac{1}{n}\right).$$

Through this last expression for S_n, we can visualize this partial sum as a sum of areas of rectangles with heights $\frac{1}{m}$ and bases of length 1, as shown in Figure 8.3.4, which uses the 9th partial sum.

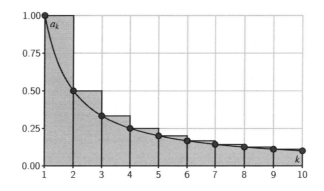

Figure 8.3.4: A picture of the 9th partial sum of the harmonic series as a sum of areas of rectangles.

The graph of the continuous function f defined by $f(x) = \frac{1}{x}$ is overlaid on this plot.

a. Explain how this picture represents a particular Riemann sum.

b. What is the definite integral that corresponds to the Riemann sum you considered in (a)?

c. Which is larger, the definite integral in (b), or the corresponding partial sum S_9 of the series? Why?

d. If instead of considering the 9th partial sum, we consider the nth partial sum, and we let n go to infinity, we can then compare the series $\sum_{k=1}^{\infty} \frac{1}{k}$ to the improper integral $\int_1^{\infty} \frac{1}{x}\, dx$. Which of these quantities is larger? Why?

e. Does the improper integral $\int_1^{\infty} \frac{1}{x}\, dx$ converge or diverge? What does that result, together with your work in (d), tell us about the series $\sum_{k=1}^{\infty} \frac{1}{k}$?

Activity 8.3.6. The series $\sum \frac{1}{k^p}$ are special series called p-series. We have already seen that the p-series with $p = 1$ (the harmonic series) diverges. We investigate the behavior of other p-series in this activity.

a. Evaluate the improper integral $\int_1^\infty \frac{1}{x^2} \, dx$. Does the series $\sum_{k=1}^\infty \frac{1}{k^2}$ converge or diverge? Explain.

b. Evaluate the improper integral $\int_1^\infty \frac{1}{x^p} \, dx$ where $p > 1$. For which values of p can we conclude that the series $\sum_{k=1}^\infty \frac{1}{k^p}$ converges?

c. Evaluate the improper integral $\int_1^\infty \frac{1}{x^p} \, dx$ where $p < 1$. What does this tell us about the corresponding p-series $\sum_{k=1}^\infty \frac{1}{k^p}$?

d. Summarize your work in this activity by completing the following statement.

   The p-series $\sum_{k=1}^\infty \frac{1}{k^p}$ converges if and only if _____.

Activity 8.3.7. Consider the series $\sum \frac{k+1}{k^3+2}$. Since the convergence or divergence of a series only depends on the behavior of the series for large values of k, we might examine the terms of this series more closely as k gets large.

a. By computing the value of $\frac{k+1}{k^3+2}$ for $k = 100$ and $k = 1000$, explain why the terms $\frac{k+1}{k^3+2}$ are essentially $\frac{k}{k^3}$ when k is large.

b. Let's formalize our observations in (a) a bit more. Let $a_k = \frac{k+1}{k^3+2}$ and $b_k = \frac{k}{k^3}$. Calculate

$$\lim_{k\to\infty} \frac{a_k}{b_k}.$$

What does the value of the limit tell you about a_k and b_k for large values of k? Compare your response from part (a).

c. Does the series $\sum \frac{k}{k^3}$ converge or diverge? Why? What do you think that tells us about the convergence or divergence of the series $\sum \frac{k+1}{k^3+2}$? Explain.

Activity 8.3.8. Use the Limit Comparison Test to determine the convergence or divergence of the series

$$\sum \frac{3k^2 + 1}{5k^4 + 2k + 2}.$$

by comparing it to the series $\sum \frac{1}{k^2}$.

Activity 8.3.9. Consider the series defined by

$$\sum_{k=1}^{\infty} \frac{2^k}{3^k - k}.$$
(8.3.1)

This series is not a geometric series, but this activity will illustrate how we might compare this series to a geometric one. Recall that a series $\sum a_k$ is geometric if the ratio $\frac{a_{k+1}}{a_k}$ is always the same. For the series in (8.3.4), note that $a_k = \frac{2^k}{3^k-k}$.

a. To see if $\sum \frac{2^k}{3^k-k}$ is comparable to a geometric series, we analyze the ratios of successive terms in the series. Complete Table 8.3.6, listing your calculations to at least 8 decimal places.

k	5	10	20	21	22	23	24	25
$\frac{a_{k+1}}{a_k}$								

Table 8.3.6: Ratios of successive terms in the series $\sum \frac{2^k}{3^k-k}$

b. Based on your calculations in Table 8.3.6, what can we say about the ratio $\frac{a_{k+1}}{a_k}$ if k is large?

c. Do you agree or disagree with the statement: "the series $\sum \frac{2^k}{3^k-k}$ is approximately geometric when k is large"? If not, why not? If so, do you think the series $\sum \frac{2^k}{3^k-k}$ converges or diverges? Explain.

Activity 8.3.10. Determine whether each of the following series converges or diverges. Explicitly state which test you use.

a. $\sum \frac{k}{2^k}$

b. $\sum \frac{k^3+2}{k^2+1}$

c. $\sum \frac{10^k}{k!}$

d. $\sum \frac{k^3-2k^2+1}{k^6+4}$

8.4 Alternating Series

Preview Activity 8.4.1. Preview Activity 8.3.1 showed how we can approximate the number e with linear, quadratic, and other polynomial approximations. We use a similar approach in this activity to obtain linear and quadratic approximations to $\ln(2)$. Along the way, we encounter a type of series that is different than most of the ones we have seen so far. Throughout this activity, let $f(x) = \ln(1 + x)$.

   a. Find the tangent line to f at $x = 0$ and use this linearization to approximate $\ln(2)$. That is, find $L(x)$, the tangent line approximation to $f(x)$, and use the fact that $L(1) \approx f(1)$ to estimate $\ln(2)$.

   b. The linearization of $\ln(1 + x)$ does not provide a very good approximation to $\ln(2)$ since 1 is not that close to 0. To obtain a better approximation, we alter our approach; instead of using a straight line to approximate $\ln(2)$, we use a quadratic function to account for the concavity of $\ln(1 + x)$ for x close to 0. With the linearization, both the function's value and slope agree with the linearization's value and slope at $x = 0$. We will now make a quadratic approximation $P_2(x)$ to $f(x) = \ln(1 + x)$ centered at $x = 0$ with the property that $P_2(0) = f(0)$, $P_2'(0) = f'(0)$, and $P_2''(0) = f''(0)$.

   i. Let $P_2(x) = x - \frac{x^2}{2}$. Show that $P_2(0) = f(0)$, $P_2'(0) = f'(0)$, and $P_2''(0) = f''(0)$. Use $P_2(x)$ to approximate $\ln(2)$ by using the fact that $P_2(1) \approx f(1)$.

   ii. We can continue approximating $\ln(2)$ with polynomials of larger degree whose derivatives agree with those of f at 0. This makes the polynomials fit the graph of f better for more values of x around 0. For example, let $P_3(x) = x - \frac{x^2}{2} + \frac{x^3}{3}$. Show that $P_3(0) = f(0)$, $P_3'(0) = f'(0)$, $P_3''(0) = f''(0)$, and $P_3'''(0) = f'''(0)$. Taking a similar approach to preceding questions, use $P_3(x)$ to approximate $\ln(2)$.

   iii. If we used a degree 4 or degree 5 polynomial to approximate $\ln(1 + x)$, what approximations of $\ln(2)$ do you think would result? Use the preceding questions to conjecture a pattern that holds, and state the degree 4 and degree 5 approximation.

Activity 8.4.2. Remember that, by definition, a series converges if and only if its corresponding sequence of partial sums converges.

   a. Calculate the first few partial sums (to 10 decimal places) of the alternating series

$$\sum_{k=1}^{\infty}(-1)^{k+1}\frac{1}{k}.$$

Label each partial sum with the notation $S_n = \sum_{k=1}^{n}(-1)^{k+1}\frac{1}{k}$ for an appropriate choice of n.

   b. Plot the sequence of partial sums from part (a). What do you notice about this sequence?

Activity 8.4.3. Which series converge and which diverge? Justify your answers.

a. $\displaystyle\sum_{k=1}^{\infty} \frac{(-1)^k}{k^2 + 2}$

b. $\displaystyle\sum_{k=1}^{\infty} \frac{(-1)^{k+1}2k}{k + 5}$

c. $\displaystyle\sum_{k=2}^{\infty} \frac{(-1)^k}{\ln(k)}$

Activity 8.4.4. Determine the number of terms it takes to approximate the sum of the convergent alternating series

$$\sum_{k=1}^{\infty} \frac{(-1)^{k+1}}{k^4}$$

to within 0.0001.

Activity 8.4.5.

a. Explain why the series

$$1 - \frac{1}{4} - \frac{1}{9} + \frac{1}{16} + \frac{1}{25} + \frac{1}{36} - \frac{1}{49} - \frac{1}{64} - \frac{1}{81} - \frac{1}{100} + \cdots$$

must have a sum that is less than the series

$$\sum_{k=1}^{\infty} \frac{1}{k^2}.$$

b. Explain why the series

$$1 - \frac{1}{4} - \frac{1}{9} + \frac{1}{16} + \frac{1}{25} + \frac{1}{36} - \frac{1}{49} - \frac{1}{64} - \frac{1}{81} - \frac{1}{100} + \cdots$$

must have a sum that is greater than the series

$$\sum_{k=1}^{\infty} -\frac{1}{k^2}.$$

c. Given that the terms in the series

$$1 - \frac{1}{4} - \frac{1}{9} + \frac{1}{16} + \frac{1}{25} + \frac{1}{36} - \frac{1}{49} - \frac{1}{64} - \frac{1}{81} - \frac{1}{100} + \cdots$$

converge to 0, what do you think the previous two results tell us about the convergence status of this series?

Activity 8.4.6.

a. Consider the series $\sum (-1)^k \frac{\ln(k)}{k}$.

  i. Does this series converge? Explain.

  ii. Does this series converge absolutely? Explain what test you use to determine your answer.

b. Consider the series $\sum (-1)^k \frac{\ln(k)}{k^2}$.

  i. Does this series converge? Explain.

  ii. Does this series converge absolutely? Hint: Use the fact that $\ln(k) < \sqrt{k}$ for large values of k and then compare to an appropriate p-series.

Activity 8.4.7. For (a)-(j), use appropriate tests to determine the convergence or divergence of the following series. Throughout, if a series is a convergent geometric series, find its sum.

a. $\displaystyle\sum_{k=3}^{\infty} \frac{2}{\sqrt{k-2}}$

b. $\displaystyle\sum_{k=1}^{\infty} \frac{k}{1+2k}$

c. $\displaystyle\sum_{k=0}^{\infty} \frac{2k^2+1}{k^3+k+1}$

d. $\displaystyle\sum_{k=0}^{\infty} \frac{100^k}{k!}$

e. $\displaystyle\sum_{k=1}^{\infty} \frac{2^k}{5^k}$

f. $\displaystyle\sum_{k=1}^{\infty} \frac{k^3-1}{k^5+1}$

g. $\displaystyle\sum_{k=2}^{\infty} \frac{3^{k-1}}{7^k}$

h. $\displaystyle\sum_{k=2}^{\infty} \frac{1}{k^k}$

i. $\displaystyle\sum_{k=1}^{\infty} \frac{(-1)^{k+1}}{\sqrt{k+1}}$

j. $\displaystyle\sum_{k=2}^{\infty} \frac{1}{k\ln(k)}$

k. Determine a value of n so that the nth partial sum S_n of the alternating series $\displaystyle\sum_{n=2}^{\infty} \frac{(-1)^n}{\ln(n)}$ approximates the sum to within 0.001.

8.5 Taylor Polynomials and Taylor Series

Preview Activity 8.5.1. Preview Activity 8.3.1 showed how we can approximate the number e using linear, quadratic, and other polynomial functions; we then used similar ideas in Preview Activity 8.4.1 to approximate $\ln(2)$. In this activity, we review and extend the process to find the "best" quadratic approximation to the exponential function e^x around the origin. Let $f(x) = e^x$ throughout this activity.

a. Find a formula for $P_1(x)$, the linearization of $f(x)$ at $x = 0$. (We label this linearization P_1 because it is a first degree polynomial approximation.) Recall that $P_1(x)$ is a good approximation to $f(x)$ for values of x close to 0. Plot f and P_1 near $x = 0$ to illustrate this fact.

b. Since $f(x) = e^x$ is not linear, the linear approximation eventually is not a very good one. To obtain better approximations, we want to develop a different approximation that "bends" to make it more closely fit the graph of f near $x = 0$. To do so, we add a quadratic term to $P_1(x)$. In other words, we let

$$P_2(x) = P_1(x) + c_2 x^2$$

for some real number c_2. We need to determine the value of c_2 that makes the graph of $P_2(x)$ best fit the graph of $f(x)$ near $x = 0$.

Remember that $P_1(x)$ was a good linear approximation to $f(x)$ near 0; this is because $P_1(0) = f(0)$ and $P_1'(0) = f'(0)$. It is therefore reasonable to seek a value of c_2 so that

$$P_2(0) = f(0), \qquad\qquad P_2'(0) = f'(0), \qquad\qquad \text{and } P_2''(0) = f''(0).$$

Remember, we are letting $P_2(x) = P_1(x) + c_2 x^2$.

   i. Calculate $P_2(0)$ to show that $P_2(0) = f(0)$.

   ii. Calculate $P_2'(0)$ to show that $P_2'(0) = f'(0)$.

   iii. Calculate $P_2''(x)$. Then find a value for c_2 so that $P_2''(0) = f''(0)$.

   iv. Explain why the condition $P_2''(0) = f''(0)$ will put an appropriate "bend" in the graph of P_2 to make P_2 fit the graph of f around $x = 0$.

Activity 8.5.2. We have just seen that the nth order Taylor polynomial centered at $a = 0$ for the exponential function e^x is

$$\sum_{k=0}^{n} \frac{x^k}{k!}.$$

In this activity, we determine small order Taylor polynomials for several other familiar functions, and look for general patterns.

   a. Let $f(x) = \frac{1}{1-x}$.

   i. Calculate the first four derivatives of $f(x)$ at $x = 0$. Then find the fourth order Taylor polynomial $P_4(x)$ for $\frac{1}{1-x}$ centered at 0.

   ii. Based on your results from part (i), determine a general formula for $f^{(k)}(0)$.

   b. Let $f(x) = \cos(x)$.

   i. Calculate the first four derivatives of $f(x)$ at $x = 0$. Then find the fourth order Taylor polynomial $P_4(x)$ for $\cos(x)$ centered at 0.

   ii. Based on your results from part (i), find a general formula for $f^{(k)}(0)$. (Think about how k being even or odd affects the value of the kth derivative.)

   c. Let $f(x) = \sin(x)$.

   i. Calculate the first four derivatives of $f(x)$ at $x = 0$. Then find the fourth order Taylor polynomial $P_4(x)$ for $\sin(x)$ centered at 0.

   ii. Based on your results from part (i), find a general formula for $f^{(k)}(0)$. (Think about how k being even or odd affects the value of the kth derivative.)

Activity 8.5.3. In Activity 8.5.2 we determined small order Taylor polynomials for a few familiar functions, and also found general patterns in the derivatives evaluated at 0. Use that information to write the Taylor series centered at 0 for the following functions.

a. $f(x) = \frac{1}{1-x}$

b. $f(x) = \cos(x)$ (You will need to carefully consider how to indicate that many of the coefficients are 0. Think about a general way to represent an even integer.)

c. $f(x) = \sin(x)$ (You will need to carefully consider how to indicate that many of the coefficients are 0. Think about a general way to represent an odd integer.)

d. $f(x) = \frac{1}{1+x}$

Activity 8.5.4.

a. Plot the graphs of several of the Taylor polynomials centered at 0 (of order at least 5) for e^x and convince yourself that these Taylor polynomials converge to e^x for every value of x.

b. Draw the graphs of several of the Taylor polynomials centered at 0 (of order at least 6) for $\cos(x)$ and convince yourself that these Taylor polynomials converge to $\cos(x)$ for every value of x. Write the Taylor series centered at 0 for $\cos(x)$.

c. Draw the graphs of several of the Taylor polynomials centered at 0 for $\frac{1}{1-x}$. Based on your graphs, for what values of x do these Taylor polynomials appear to converge to $\frac{1}{1-x}$? How is this situation different from what we observe with e^x and $\cos(x)$? In addition, write the Taylor series centered at 0 for $\frac{1}{1-x}$.

Activity 8.5.5.

a. Use the Ratio Test to explicitly determine the interval of convergence of the Taylor series for $f(x) = \frac{1}{1-x}$ centered at $x = 0$.

b. Use the Ratio Test to explicitly determine the interval of convergence of the Taylor series for $f(x) = \cos(x)$ centered at $x = 0$.

c. Use the Ratio Test to explicitly determine the interval of convergence of the Taylor series for $f(x) = \sin(x)$ centered at $x = 0$.

Activity 8.5.6. Let $P_n(x)$ be the nth order Taylor polynomial for $\sin(x)$ centered at $x = 0$. Determine how large we need to choose n so that $P_n(2)$ approximates $\sin(2)$ to 20 decimal places.

Activity 8.5.7.

a. Show that the Taylor series centered at 0 for $\cos(x)$ converges to $\cos(x)$ for every real number x.

b. Next we consider the Taylor series for e^x.

    i. Show that the Taylor series centered at 0 for e^x converges to e^x for every nonnegative value of x.

    ii. Show that the Taylor series centered at 0 for e^x converges to e^x for every negative value of x.

    iii. Explain why the Taylor series centered at 0 for e^x converges to e^x for every real number x. Recall that we earlier showed that the Taylor series centered at 0 for e^x converges for all x, and we have now completed the argument that the Taylor series for e^x actually converges to e^x for all x.

c. Let $P_n(x)$ be the nth order Taylor polynomial for e^x centered at 0. Find a value of n so that $P_n(5)$ approximates e^5 correct to 8 decimal places.

8.6 Power Series

Preview Activity 8.6.1. In Chapter 7, we learned some of the many important applications of differential equations, and learned some approaches to solve or analyze them. Here, we consider an important approach that will allow us to solve a wider variety of differential equations.

Let's consider the familiar differential equation from exponential population growth given by

$$y' = ky, \tag{8.6.1}$$

where k is the constant of proportionality. While we can solve this differential equation using methods we have already learned, we take a different approach now that can be applied to a much larger set of differential equations. For the rest of this activity, let's assume that $k = 1$. We will use our knowledge of Taylor series to find a solution to the differential equation (8.6.1).

To do so, we assume that we have a solution $y = f(x)$ and that $f(x)$ has a Taylor series that can be written in the form

$$y = f(x) = \sum_{k=0}^{\infty} a_k x^k,$$

where the coefficients a_k are undetermined. Our task is to find the coefficients.

a. Assume that we can differentiate a power series term by term. By taking the derivative of $f(x)$ with respect to x and substituting the result into the differential equation (8.6.1), show that the equation

$$\sum_{k=1}^{\infty} k a_k x^{k-1} = \sum_{k=0}^{\infty} a_k x^k$$

must be satisfied in order for $f(x) = \sum_{k=0}^{\infty} a_k x^k$ to be a solution of the DE.

b. Two series are equal if and only if they have the same coefficients on like power terms. Use this fact to find a relationship between a_1 and a_0.

c. Now write a_2 in terms of a_1. Then write a_2 in terms of a_0.

d. Write a_3 in terms of a_2. Then write a_3 in terms of a_0.

e. Write a_4 in terms of a_3. Then write a_4 in terms of a_0.

f. Observe that there is a pattern in (b)-(e). Find a general formula for a_k in terms of a_0.

g. Write the series expansion for y using only the unknown coefficient a_0. From this, determine what familiar functions satisfy the differential equation (8.6.1). (*Hint*: Compare to a familiar Taylor series.)

Activity 8.6.2. Determine the interval of convergence of each power series.

a. $\sum_{k=1}^{\infty} \frac{(x-1)^k}{3k}$

b. $\sum_{k=1}^{\infty} kx^k$

c. $\sum_{k=1}^{\infty} \frac{k^2(x+1)^k}{4^k}$

d. $\sum_{k=1}^{\infty} \frac{x^k}{(2k)!}$

e. $\sum_{k=1}^{\infty} k!x^k$

Activity 8.6.3. Our goal in this activity is to find a power series expansion for $f(x) = \frac{1}{1+x^2}$ centered at $x = 0$.

While we could use the methods of Section 8.5 and differentiate $f(x) = \frac{1}{1+x^2}$ several times to look for patterns and find the Taylor series for $f(x)$, we seek an alternate approach because of how complicated the derivatives of $f(x)$ quickly become.

    a. What is the Taylor series expansion for $g(x) = \frac{1}{1-x}$? What is the interval of convergence of this series?

    b. How is $g(-x^2)$ related to $f(x)$? Explain, and hence substitute $-x^2$ for x in the power series expansion for $g(x)$. Given the relationship between $g(-x^2)$ and $f(x)$, how is the resulting series related to $f(x)$?

    c. For which values of x will this power series expansion for $f(x)$ be valid? Why?

Activity 8.6.4. Let f be the function given by the power series expansion

$$f(x) = \sum_{k=0}^{\infty} (-1)^k \frac{x^{2k}}{(2k)!}.$$

a. Assume that we can differentiate a power series term by term, just like we can differentiate a (finite) polynomial. Use the fact that

$$f(x) = 1 - \frac{x^2}{2!} + \frac{x^4}{4!} - \frac{x^6}{6!} + \cdots + (-1)^k \frac{x^{2k}}{(2k)!} + \cdots$$

to find a power series expansion for $f'(x)$.

b. Observe that $f(x)$ and $f'(x)$ have familiar Taylor series. What familiar functions are these? What known relationship does our work demonstrate?

c. What is the series expansion for $f''(x)$? What familiar function is $f''(x)$?

Activity 8.6.5. Find a power series expansion for $\ln(1 + x)$ centered at $x = 0$ and determine its interval of convergence.

Colophon

This book was authored in PreTeXt.

Made in the USA
Las Vegas, NV
04 October 2022

56526767R00114